Color
Rendering
Capacity
of Light

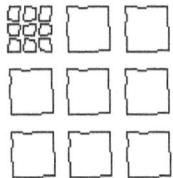

Color Rendering Capacity of Light

A MONOGRAPH

Xu He (H.Xu)

I dedicate this book to the memory of my mother and father, Xu Hancheng and Xu Jing, to my sisters and brother, to my wife and daughter, and to each friend of mine. It was with their natural kindness accompanying all the way that I developed the idea introduced in this book.

Great thanks also go to the members of Lulu. It is their imagination and dedication that finally lead to the birth of this book.

© 2010 by Xu He

Published in the United States by Lulu, Raleigh, www.lulu.com

Color Rendering Capacity of Light/ Xu He

ISBN 978-0-557-29359-9

1. Color rendering – Science. [1. Color rendering.] I. Title

Contents

1 Introduction

1.1 Color Rendering of Light

A light source such as the flame of a candle or the sun gives out light continuously. While traveling forward, the light may encounter a surface on its way and be reflected from the surface. If the light finally enters the eyes of an observer, it would evoke a sensation called "color" on the part of the observer. Usually, the observer would ascribe this sensation of color to the surface reflecting the light, and think that she or he sees a colored object in front.

On leaving the light source, the light is carrying color characteristics of the light source. And after being reflected from the surface, the light is furthermore carrying color characteristics of the surface reflecting the light. So the color as sensed finally by the observer depends on what the illuminating light source is, as well as what the reflecting surface is. Anyway, it is with the help of the light that enables us to see a colorful world.

Under a given light source, different kinds of surfaces would appear in different colors. On the other hand, the very same surface when illuminated by different kinds of light sources could also appear in different colors. In other words, different kinds of light could render colors in different ways. Thus, with a different kind of light, we could see a different colorful world. This is the subject this book intends to discuss, the color rendering of light.

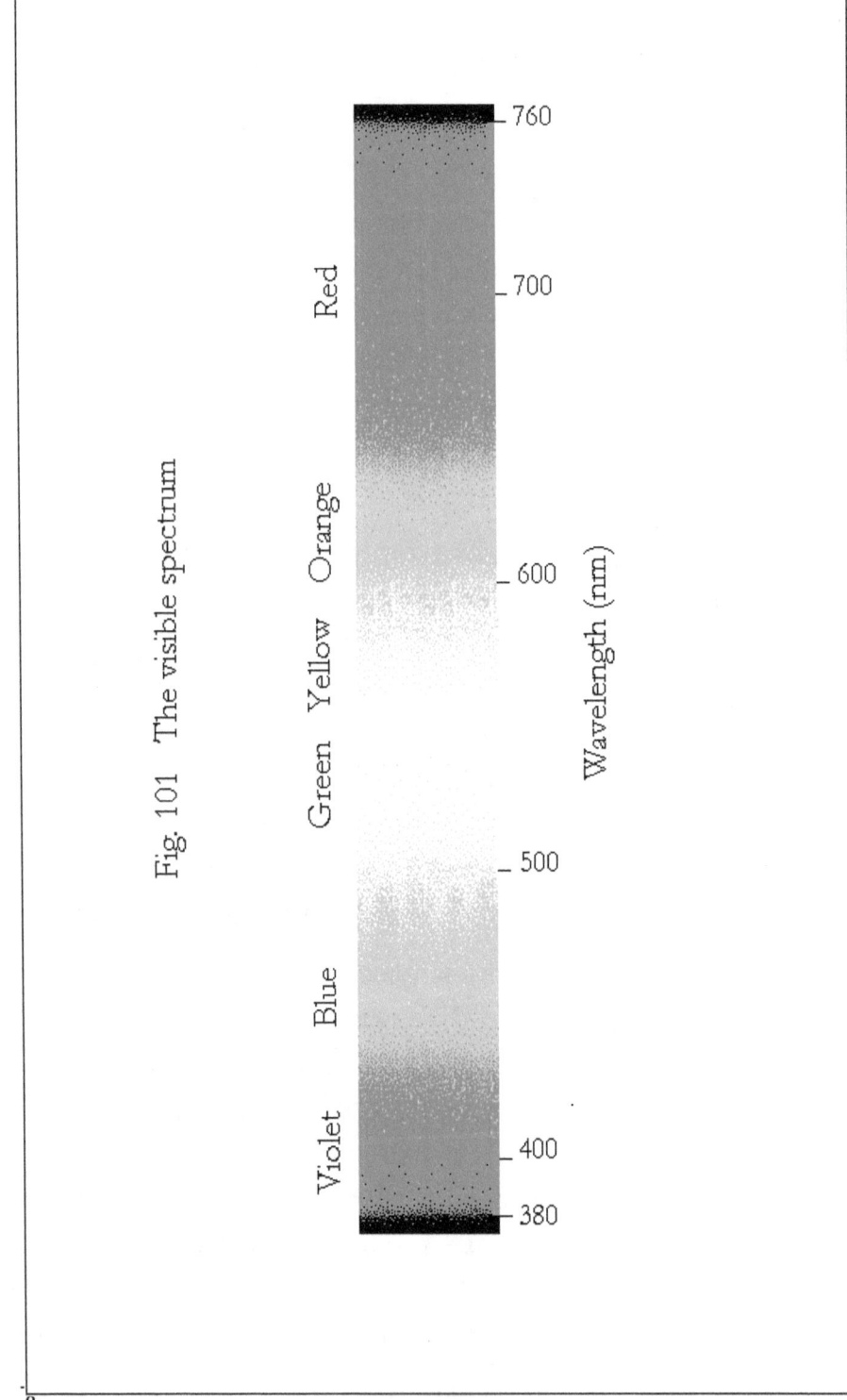

Fig. 101 The visible spectrum

1.2 Spectral Power Distributions

Since scientist Isaac Newton led a beam of sunlight through a prism and saw behind the prism a spread colorful band, i.e. a spectrum of colors, people had understood that light could be regarded as electromagnetic radiation within a certain wavelength range. The visible spectrum covers a wavelength range from about 380 nm to 760 nm, see Fig. 101. The light of a single wavelength, say 400 nm, could evoke a sensation of pure blue, and the light of 700 nm a sensation of pure red. The color evoked by the light of a single wavelength is called a spectrum color.

The light from any actual light source is composed of a series of spectral components mixed in certain proportions. Each kind of light has a particular kind of spectral composition, and can be specified by a particular spectral power distribution $P(\lambda)$. Shown in Fig. 102 are the spectral power distributions with several kinds of light sources: CIE illuminants A and D65, high pressure sodium, and tri-band fluorescent. CIE illuminant A is similar to an incandescent tungsten lamp having the spectral power distribution of a blackbody radiator at a temperature of 2856K. CIE illuminant D65 is actually a set of data representing the spectral power distribution of average daylight. The data for these light sources are listed in Appendix 1.

Consider two beams of light, one beam from one candle, the other from two candles. Suppose the

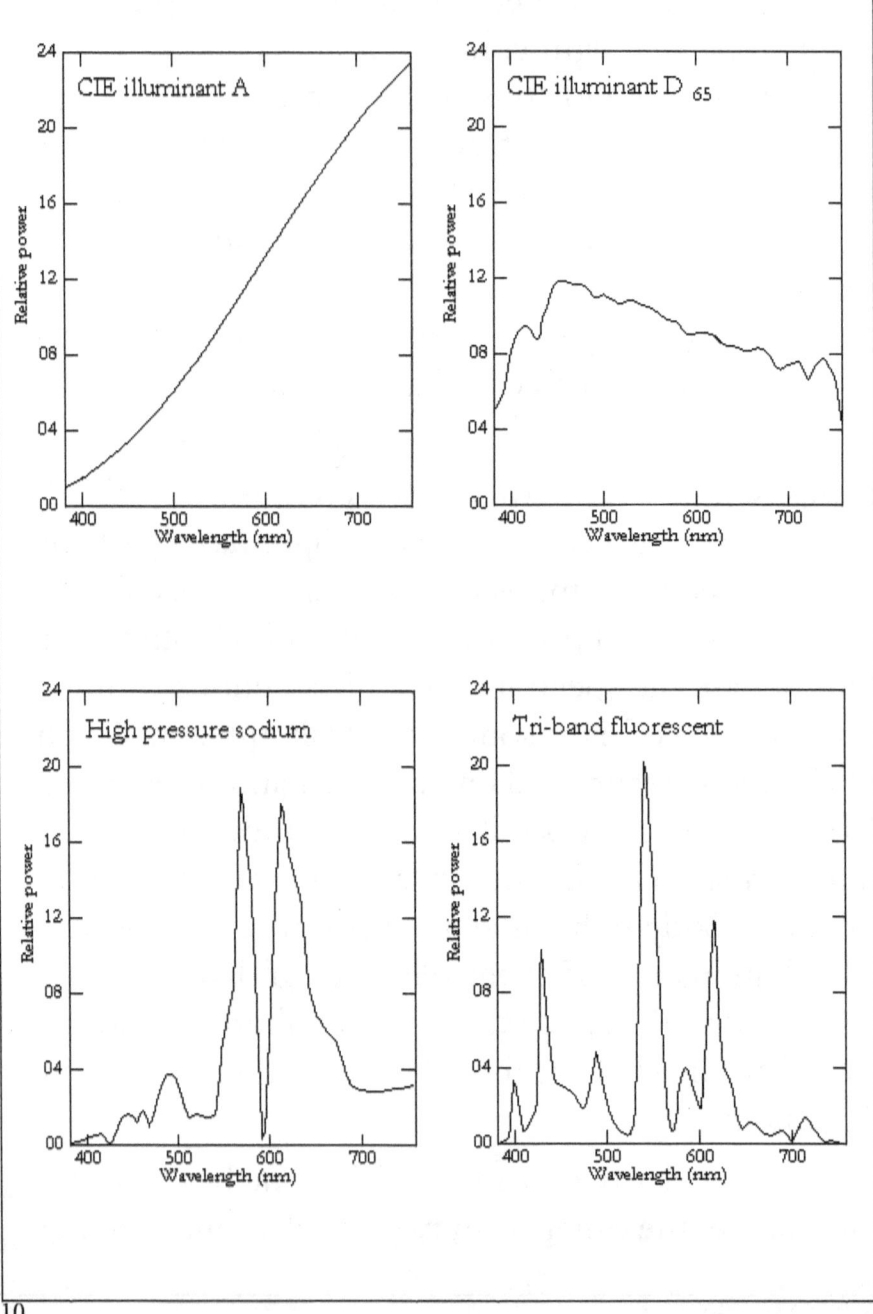

Fig. 102 Spectral power distributions of several kinds of light sources

spectral power distribution in the former is P(λ), the spectral power distribution in the latter will be 2P(λ). These two beams come from the same "kind" of light sources, have the same kind of spectral composition, i.e. the same "relative" spectral power distribution. The only difference between them is that the amount of light with one of them is twice as much as that with the other.

1.3 Spectral Reflectance Curves

When a surface is illuminated by a certain light source, part of the light falling on the surface will be reflected from the surface. The amount of the light reflected from the surface is never larger than the amount of the incident light. The ratio of the amount of light reflected to the amount of light incident is called the reflectance of the surface, which has a value between 0.0 and 1.0.

A detailed examination shows that the surface is reflecting each spectral component of the incident light in a particular proportion that depends on one of the inherent properties of the surface. And this inherent property can be described by a spectral reflectance curve R(λ). Different sorts of surfaces have different spectral reflectance curves.

Suppose the spectral power distribution of a given light source is P(λ), the spectral reflectance curve of a given surface is R(λ), and this surface is illuminated by that light source, then the spectral

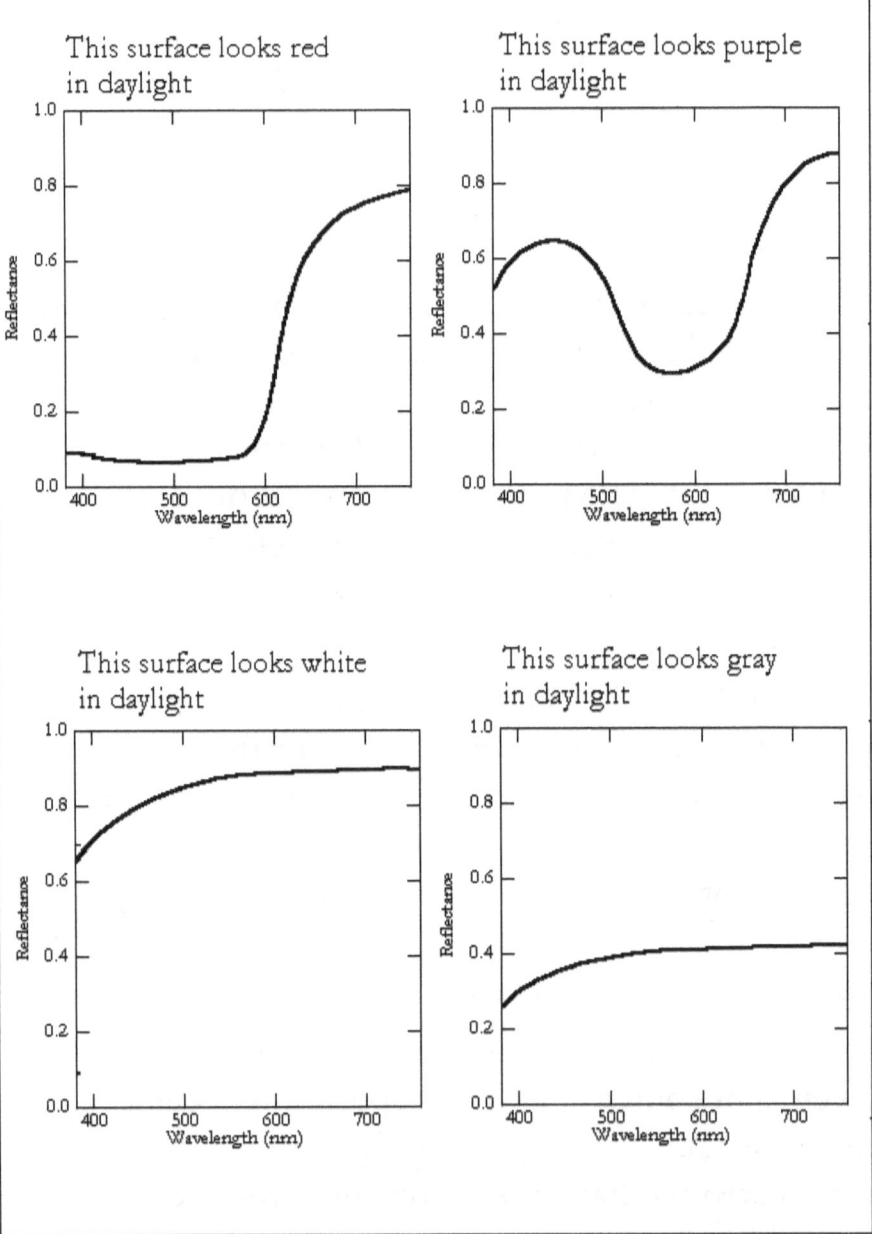

Fig. 103   Spectral reflectance curves of several surfaces

power distribution of the light reflected from this surface will be $P(\lambda)R(\lambda)$. When the reflected light enters the eyes of an observer, it will evoke a sensation of color, which is the color of the given surface under the given light source.

The spectral reflectance curves with several sorts of surfaces are shown in Fig. 103.

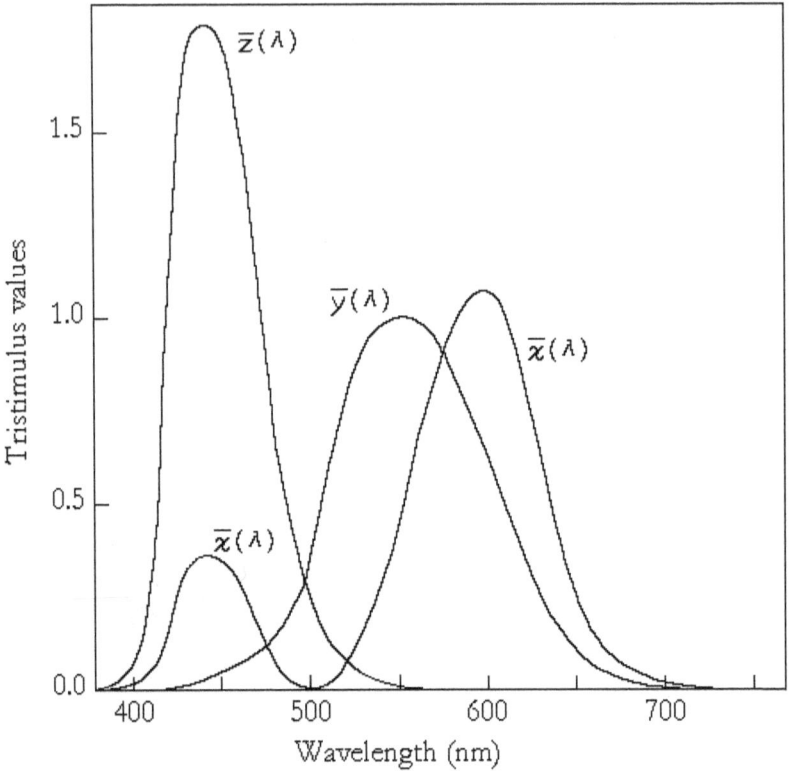

Fig. 201   CIE 1931 standard observer color-matching functions $\bar{x}(\lambda)$, $\bar{y}(\lambda)$, and $\bar{z}(\lambda)$, i.e. tristimulus values of spectrum colors

2 Numerical Expression of Color

2.1 Color-Matching Functions

Color specification in terms of equivalent stimuli had been adopted by the International Commission on Illumination, i.e. Commission Internationale de l'Eclairage (CIE). The method is based on the fact that a normal observer can duplicate the effect of any color stimulus by combining the light from three primary stimuli in the proper proportions.

On the basis of the average of a series of measurements with carefully selected observers and the extensive investigation by many scientists, the CIE in 1931 recommended a set of data called the CIE standard observer color-matching functions $\underline{x}(\lambda)$, $\underline{y}(\lambda)$, and $\underline{z}(\lambda)$. The values of $\underline{x}(\lambda)$, $\underline{y}(\lambda)$, and $\underline{z}(\lambda)$ indicate the amount of each of the CIE chosen primaries that is required by a normal observer to make their combination match the spectrum color of one watt of the indicated wavelength. The color-matching functions can also be called the tristimulus values of the spectrum colors, and are represented graphically in Fig. 201. The CIE 1931 color-matching data at 1nm wavelength intervals are given in Appendix 2.

2.2 Colors of Light Sources

The light source color refers to a color evoked by the light coming directly from a certain light source. The color evoked by the light from the light source of

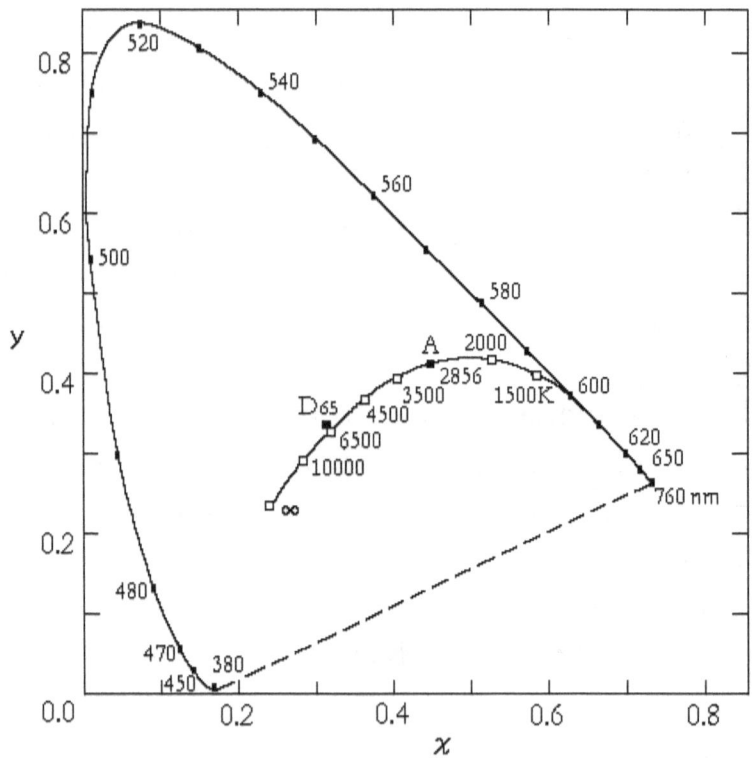

Fig. 202   CIE 1931 chromaticity diagram with the locus of the spectrum colors, the locus of the chromaticities of blackbody radiators at various temperatures (□), and the locations of CIE illuminants (■) A and D65 being shown

spectral power distribution $P(\lambda)$ can be specified by the following tristimulus values X, Y, and Z:

$X=\Sigma\ P(\lambda)\underline{x}(\lambda)$

$=P(380)\underline{x}(380)+P(390)\underline{x}(390)+ \ldots +P(760)\underline{x}(760)$.

In case higher precision is required, calculation should be carried out with data at 1nm wavelength intervals, instead of 10nm wavelength intervals, that is:

$X=\Sigma\ P(\lambda)\underline{x}(\lambda)$

$=P(380)\underline{x}(380)+P(381)\underline{x}(381)+ \ldots +P(769)\underline{x}(769)$.

Similarly,

$Y=\Sigma\ P(\lambda)\underline{y}(\lambda)$.

$Z=\Sigma\ P(\lambda)\underline{z}(\lambda)$.

Among X, Y, and Z, tristimulus value Y has a plain meaning: the value of Y is proportional to the amount, i.e. the luminous flux, of the light evoking the color. In fact, if a proper unit is adopted, the value of Y is just equal to the luminous flux.

In practice, the color is often specified in terms of Y, x, and y, where x and y are called the chromaticity coordinates, or the chromaticity, of the color, and defined as follows:

$x=X/(X+Y+Z)$.

$y=Y/(X+Y+Z)$.

The chromaticity of a color can be represented by plotting the chromaticity coordinates x and y on a right-angled coordinate plane, that is the CIE 1931 chromaticity diagram, see Fig. 202. Shown on the diagram is the locus constituted from the chromaticity points of the spectrum colors. All realizable colors are represented by points within the region enclosed by the locus. This region is called the gamut of chromaticities of real colors. Also shown on the diagram is the locus of the chromaticities of blackbody radiators at various temperatures and the locations of the CIE illuminants A and D65.

It is noticeable that an incandescent body at a higher temperature, say 5000 degree centigrade rather than 1000 degree centigrade, looks blue, and the color of blue could evoke a feeling of cool psychologically, while an ordinary object at a higher temperature, say 50 degree centigrade rather than 10 degree centigrade, could evoke a feeling of warm physically.

Consider two beams of light, one beam from one candle, the other from two candles. The two colors evoked by these two beams respectively will be represented by the same point on the chromaticity diagram. That is, these two colors have the same chromaticity since they come from the same kind of light sources, have the same relative spectral power distribution.. The only difference between these two colors is that the value of Y with one of them is twice as much as that with the other.

It should be noticed that the colors evoked by some of different relative spectral power distributions can also have the same chromaticity, that is, one point on the chromaticity diagram can correspond to more than one spectral power distribution. Even if two light sources look the same in color, there is possibility for them to have different relative spectral power distributions.

2.3 Colors of Objects

The object color refers to a color evoked by the light reflected from a surface illuminated by a certain light source. Suppose $R(\lambda)$ is the spectral reflectance curve of the reflecting surface, $P(\lambda)$ is the spectral power distribution of the illuminating light source, the object color evoked by the light reflected from the given surface under the given light source is specified by the tristimulus values X, Y, and Z as follows:

$$X=\Sigma\ P(\lambda)R(\lambda)\underline{x}(\lambda)$$
$$Y=\Sigma\ P(\lambda)R(\lambda)\underline{y}(\lambda)$$
$$Z=\Sigma\ P(\lambda)R(\lambda)\underline{z}(\lambda),$$

where Y is the luminous flux of the reflected light.

If Y_0 denotes $\Sigma\ P(\lambda)\underline{y}(\lambda)$, i.e. the luminous flux of the illuminating light, then

$$Y=Y(Y_0/Y_0)=(Y/Y_0)Y_0,$$

where Y/Y_0 is the ratio of the reflected luminous flux to the illuminating luminous flux, and is called the luminous flux reflectance of the surface under the given light source.

When the illumination is fixed to one unit luminous flux, that is, if in every case, the equation

$Y_0=\Sigma\ P(\lambda)\underline{y}(\lambda)=1.0$ holds true, then the value of Y is equal to the luminous flux reflectance, which has a limited value between 0.0 and 1.0. One important point to be kept in mind!

The object color can also be specified in terms of Y, x, and y, where x and y are the chromaticity coordinates, or chromaticity, of the object color, and defined as follows:

$x=X/(X+Y+Z)$

$y=Y/(X+Y+Z)$.

In practice, x and y can also be called the chromaticity of the reflecting surface under the illuminating light.

Different sorts of surfaces illuminated by the same light source usually give different chromaticities. However, some of different sorts of surfaces under the same light source can give the same chromaticity as well. As a simplest example, the two colors evoked by the two surfaces, having spectral reflectance curves $R(\lambda)$ and $0.5R(\lambda)$ respectively, under the same light source will have the same chromaticity. The only difference between these two colors is that they have different values of Y, or different values of luminous flux reflectance. The former has a value of luminous flux reflectance twice as much as that the latter has. State the comment in other words: under a given light source, there can be more than one spectral reflectance curves which give the same chromaticity, while their values of luminous flux reflectance may and may not be identical. One more point to be kept in mind!

3 CIE Color Rendering Index

Some daylight fluorescent lamps have the same chromaticity as daylight has, but a piece of bright colored fabric selected in daylight may turn dull-looking right after brought indoors under these fluorescent lamps. Different kinds of light sources can make the very same object appear in different colors. This is because of the fact that these light sources have different kinds of spectral composition; although they can have the same chromaticity, they render colors in different ways.

To assess how closely the way a given light source renders colors resemble that a reference illuminant does, the CIE has defined a Color Rendering Index to be an average measure of how closely the color appearances of several selected color samples under the light source to be tested resemble those of the same samples under a reference illuminant. Exact match of all the color appearances is defined as a color rendering index of 100.

The spectral reflectance data of 8 color samples selected by CIE for use in the calculation of color rendering index are listed in Appendix 3

The reference illuminant usually is a blackbody radiator at a certain temperature at which the chromaticity of the blackbody is equal or close to that of the test source.

The calculation procedure for the CIE color rendering index is outlined as follows:

Fig. 301 Color differences of 8 color samples under referende source and test source

- ■ Chromaticity W*ri, U*ri, V*ri of color sample numbered i under reference source

- □ Chromaticity W*ti, U*ti, V*ti of color sample numbered i under test source

- ■⃗□ Color difference △Ei of color sample numbered i in CIE 1964 uniform color space

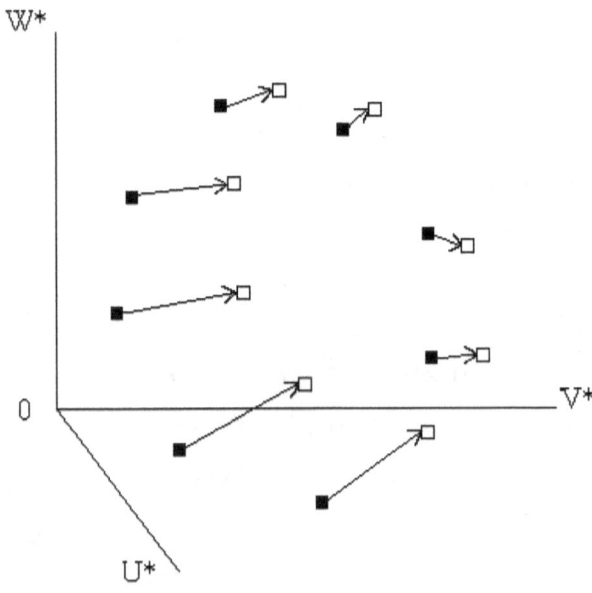

Suppose the tristimulus values for the test source and the reference illuminant are X_t, Y_t, Z_t, and X_r, Y_r, Z_r respectively. The tristimulus values for the 8 color samples under the test source and the reference illuminant respectively are X_{ti}, Y_{ti}, Z_{ti}, and X_{ri}, Y_{ri}, Z_{ri}, where i=1 to 8.

Calculate the CIE 1960 UCS chromaticity coordinates u and v as follows:

$u_t = 4X_t/(X_t+15Y_t+3Z_t)$, $v_t = 6Y_t/(X_t+15Y_t+3Z_t)$

$u_r = 4X_r/(X_r+15Y_r+3Z_r)$, $v_r = 6Y_r/(X_r+15Y_r+3Z_r)$

$u_{ti} = 4X_{ti}/(X_{ti}+15Y_{ti}+3Z_{ti})$, $v_{ti} = 6Y_{ti}/(X_{ti}+15Y_{ti}+3Z_{ti})$

$u_{ri} = 4X_{ri}/(X_{ri}+15Y_{ri}+3Z_{ri})$, $v_{ri} = 6Y_{ri}/(X_{ri}+15Y_{ri}+3Z_{ri})$

Suppose ΔEi denotes the color difference of each color sample under the test source and the reference illuminant, it is calculated as follows:

$W_{ti}^* = 25(Y_{ti})^{1/3}-17$

$U_{ti}^* = 13W_{ti}^*(u_{ti}-u_t)$

$V_{ti}^* = 13W_{ti}^*(v_{ti}-v_t)$

$W_{ri}^* = 25(Y_{ri})^{1/3}-17$

$U_{ri}^* = 13W_{ri}^*(u_{ri}-u_t)$

$V_{ri}^* = 13W_{ri}^*(v_{ri}-v_t)$

$\Delta Ei = ((U_{ri}^*-U_{ti}^*)^2+(V_{ri}^*-V_{ti}^*)^2+(W_{ri}^*-W_{ti}^*)^2)^{1/2}$

These color differences are shown in Fig. 301. The average of color differences for the 8 color samples is:

$\Delta Ea = (\Sigma\Delta Ei)/8$, (i=1 to 8)

The CIE color rendering index Ra = 100 - 4.6ΔEa

A scale coefficient of 4.6 is used to make the Ra for the standard fluorescent lamps equal to exactly 50.

In case the chromaticity of the reference illuminant is not close enough to that of the test source, CIE recommends the effect of chromatic adaptation be taken into account in the calculation of the color rendering index, see Reference 2.

4 Color Rendering Capacity

"Is it true that every kind of light can vividly render the same number of colors?" If not, then "What kind of light can render the greatest number of colors and make a scene look the most colorful?" "How many colors on earth can be rendered by a given kind of light?" Such questions highlight a particular aspect of the color rendering quality of light, an aspect relating to the concern of whether a given light source is capable of rendering a great number of widely different colors.

To find an answer to the question of how many colors on earth can be rendered by a given light source, a direct approach might be to calculate the number of all the object colors producible under the given light source. Specifically, use each of all spectral reflectance curves possible in physics, in conjunction with the spectral power distribution of the given light source, and calculate one chromaticity. Collect all the calculated chromaticities that are not identical with each other to obtain the number needed.

Since the number of all possible spectral reflectance curves is infinite, the amount of the calculation involved will be infinite and cannot be completed in any definite period of time. A different approach to the problem has to be found.

4.1 Maximum Visual Efficiency

An investigation by Prof. David L. MacAdam on visual

Fig. 401   Spectral reflectance curves of the type required to ensure the maximum visual efficiency, i.e. the maximum luminous flux reflectance

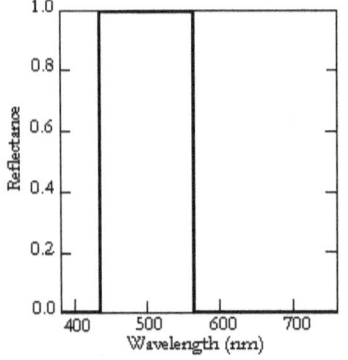
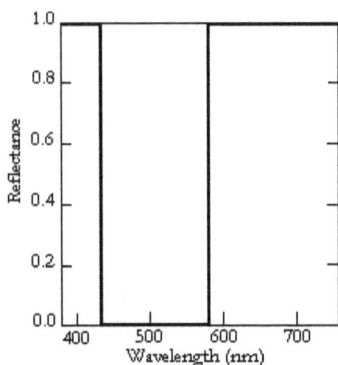

efficiency of colored materials showed that for any specified chromaticity, the maximum possible visual efficiency under a certain illuminant requires spectral reflectance curves of a particular type such that they contain exclusively either one reflecting band or one absorbing band within the visible spectrum, and that the reflectance must not be anything other than 0.0 and 1.0, see Reference 5. Such spectral reflectance curves are illustrated in Fig. 401, and will be called "square spectral reflectance curves" for short.

It should be noted that the term "visual efficiency of colored materials" mentioned above in fact is the luminous flux reflectance of the reflecting materials illuminated by a certain light source.

Now the investigation result presented above can be restated as follows:

--- Among all the spectral reflectance curves that can produce a specified chromaticity under illumination of a certain spectral power distribution, the square spectral reflectance curve gives the maximum luminous flux reflectance achievable at that chromaticity under that illumination.

Using the square spectral reflectance curves, in conjunction with a certain spectral power distribution, say the CIE illuminant D65, a series of chromaticity coordinates having a specified value of luminous flux reflectance, say 0.1, can be calculated. Plot these chromaticity coordinates on an x and y chromaticity plane, and a closed locus can be constructed from

Fig. 402   Using square spectral reflectance curves in calculation of chromaticities

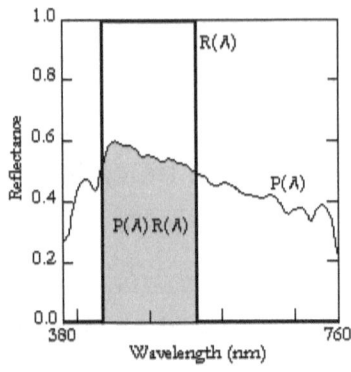 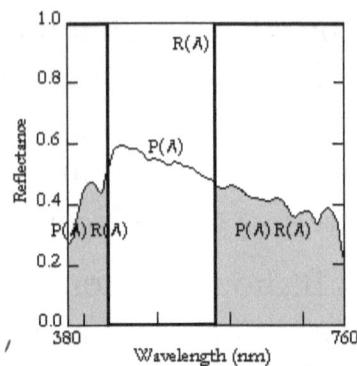

Luminous flux reflectance =

$$= \frac{\Sigma\, P(\lambda)\, R(\lambda)\, \bar{y}(\lambda)}{\Sigma\, P(\lambda)\, \bar{y}(\lambda)}$$

$X = \Sigma\, P(\lambda)\, R(\lambda)\, \bar{x}(\lambda)$
$Y = \Sigma\, P(\lambda)\, R(\lambda)\, \bar{y}(\lambda)$
$Z = \Sigma\, P(\lambda)\, R(\lambda)\, \bar{z}(\lambda)$

Calculate chromaticity point #1

#2

.
.
.

#m

Luminous flux reflectance =

$$= \frac{\Sigma\, P(\lambda)\, R(\lambda)\, \bar{y}(\lambda)}{\Sigma\, P(\lambda)\, \bar{y}(\lambda)}$$

$X = \Sigma\, P(\lambda)\, R(\lambda)\, \bar{x}(\lambda)$
$Y = \Sigma\, P(\lambda)\, R(\lambda)\, \bar{y}(\lambda)$
$Z = \Sigma\, P(\lambda)\, R(\lambda)\, \bar{z}(\lambda)$

Calculate chromaticity point #m+1

#m+2

.
.
.

#n

these chromaticity points. This locus will be called locus1, which is associated with a luminous flux reflectance of 0.1, see Fig. 402 and Fig. 403.

Any point on locus1 represents a chromaticity that has the specified value of luminous flux reflectance, i.e. 0.1, under the light source of given spectral power distribution. And this value of luminous flux reflectance, 0.1, is the maximum luminous flux reflectance achievable at that chromaticity location under the given light source, since this locus is composed of those chromaticity points produced from the square spectral reflectance curves.

Repeat the above process and plot locus2, locus3, ..., and locus9 successively. These loci are associated with luminous flux reflectance of 0.2, 0.3, ..., and 0.9 respectively, see Fig. 404.

Locus0 can also be plotted, that is the locus of the spectrum colors, being associated with a luminous flux reflectance of 0.0. There otherwise would have been a locus10, this locus has been reduced to a single point, corresponding to the chromaticity of the light source itself, being associated with a luminous flux reflectance of 1.0.

4.2 Maximum Chromaticity Range

Carefully examine locus2 in relation to its closest neighbors locus1 and locus3, see Fig. 405:

Fig. 403 Plotting a locus, locus1, from chromaticity points having a certain maximum luminous flux reflectance, 0.1

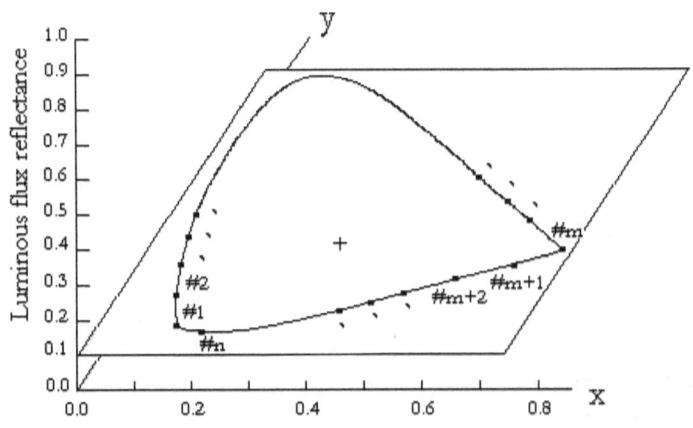

Fig. 404   Plotting locus0, locus1, ..., and locus9 associated
with maximum luminous flux reflectance of 0.0, 0.1, ...,
and 0.9 respectinely

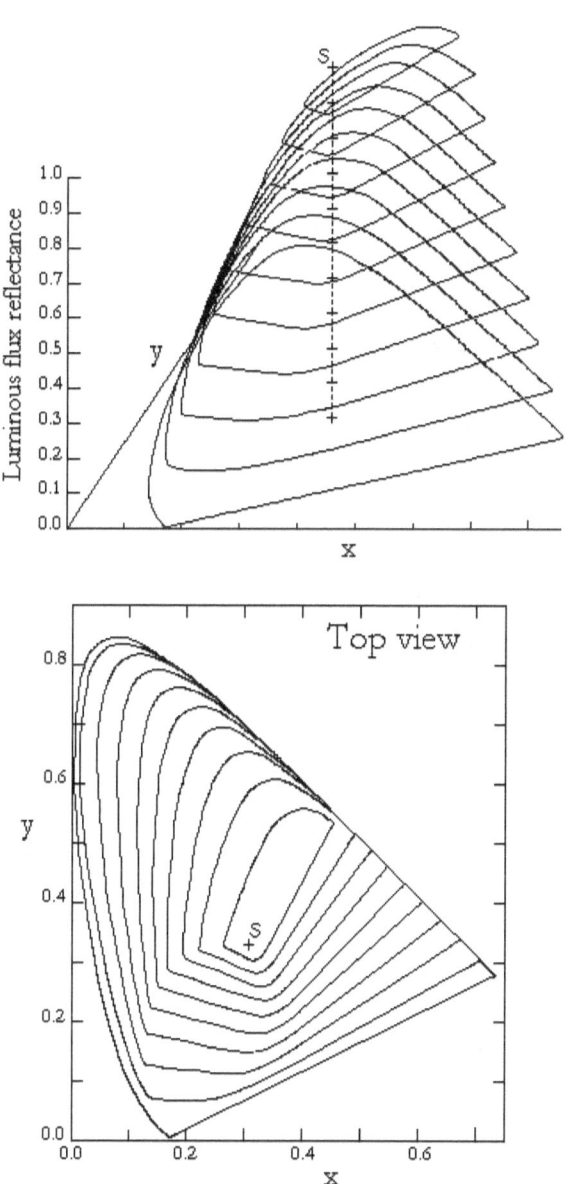

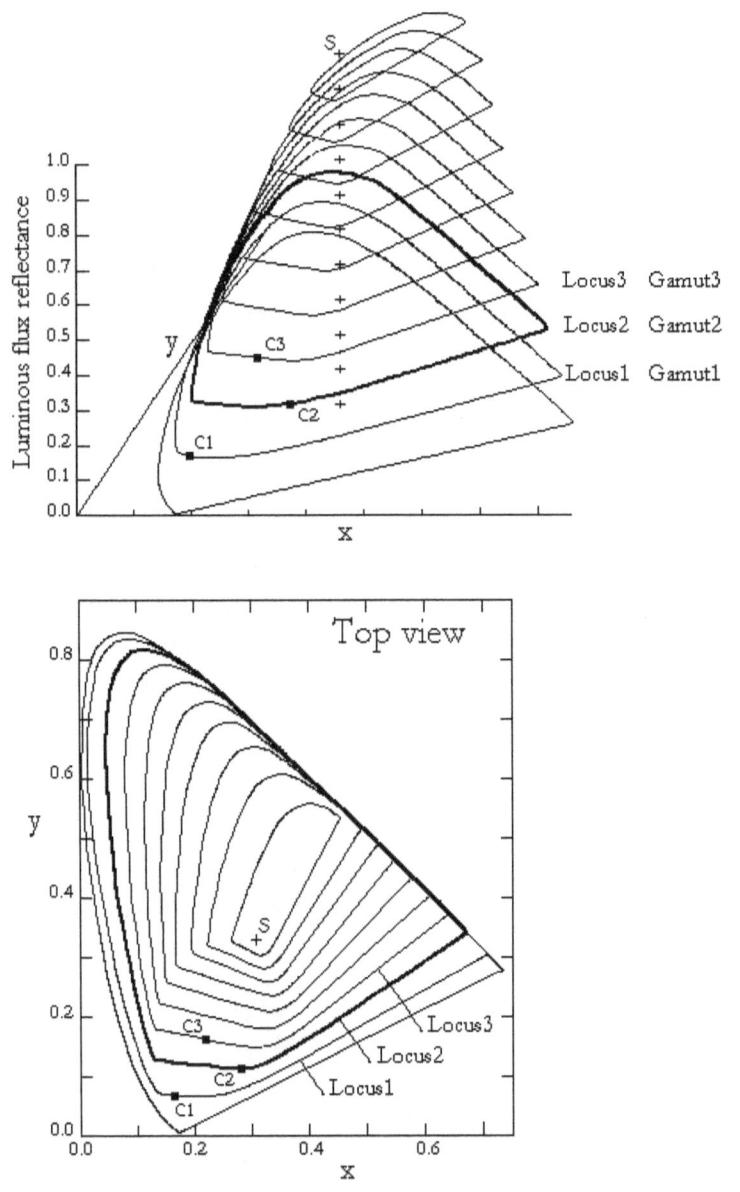

Fig. 405 The maximun chromaticity range associated with a luminous flux reflectance of 0.2, that is the region enclosed by locus2, or the gamut2

a) Any point C2 on locus2 represents a chromaticity having a luminous flux reflectance of 0.2, which is also the maximum luminous flux reflectance achievable at this chromaticity location under the light source of given spectral power distribution;

b) Any point C3 on locus3 is inside the region enclosed by locus2, and has a luminous flux reflectance higher than what C2 can have. A luminous flux reflectance of 0.3 is the maximum luminous flux reflectance achievable at the location of C3, but, of course, any luminous flux reflectance lower than 0.3, say 0.2, is also achievable at this chromaticity location;

c) Any point C1 on locus1 is outside the region enclosed by locus2. The maximum luminous flux reflectance achievable at the location of C1 is 0.1, which is lower than what C2 can have.

Putting together comments a), b), and c), an important conclusion can be reached as follows:

--- The locus2 associated with a luminous flux reflectance of 0.2 defines a region that is the Maximum Chromaticity Range, or "gamut", in which the chromaticity points having a luminous flux reflectance of 0.2 under the light source of given spectral power distribution can reside. In other words, the region enclosed by locus2 comprises all the chromaticity points having a luminous flux reflectance of 0.2, producible under the light source of given spectral power distribution; any chromaticity point outside this region can only have a luminous flux reflectance lower than 0.2 under the given light source. The region enclosed by locus2 and associated with a luminous flux reflectance of 0.2 will be called gamut2, see Fig. 405. The same argument applies to locus1 and gamut1, locus2 and gamut2, …, and locus9 and gamut9.

Considering all these gamuts put together, a further conclusion can be reached as follows:

--- A collection of all the gamuts, each being associated with one particular value of luminous flux reflectance between 0.0 and 1.0, comprises all the chromaticity points, having every value of luminous flux reflectance between 0.0 and 1.0, producible under the light source of given spectral power distribution.

4.3 Color Solid Volume Signifies Color Rendering Capacity

As shown in Section 2.3, the object color can be specified in terms of Y, x, and y, where Y can be expressed as follows:

$$Y = (Y/Y_0)Y_0$$
$$= (\Sigma P(\lambda)R(\lambda)\underline{y}(\lambda)/\Sigma P(\lambda)\underline{y}(\lambda))\Sigma P(\lambda)\underline{y}(\lambda)$$

When the illumination is fixed to one unit luminous flux, $Y = \Sigma P(\lambda)R(\lambda)\underline{y}(\lambda)/\Sigma P(\lambda)\underline{y}(\lambda)$, i.e. Y is the same as the luminous flux reflectance of the reflecting surface under the illuminating light source, and has a value limited between 0.0 and 1.0.

Thus, under the condition of the illumination being fixed to one unit luminous flux, the luminous flux reflectance in Fig.405 can be substituted with Y, as shown in Fig. 406, and the conclusion in Section 4.2 can be restated as follows:

--- A collection of all the gamuts, each being associated with one particular value of Y between 0.0 and 1.0, comprises all of the object colors producible under the given light source with one unit luminous flux. In other words, all of these gamuts piled up one by one constitutes a "color solid", within whose boundary are all the object colors producible under the given kind of light sources with one unit luminous flux.

While the number of these object colors is infinite, all of these colors are comprised within the color solid of a finite volume. And the larger the volume of the color solid in a uniform color space, the wider the range of the colors involved. These observations lead to a further conclusion as follows:

--- The volume of the color solid associated with the spectral power distribution of given light source determines the upper limit for the range of the object colors producible under the given light source with one unit luminous flux, and is proportional to the maximum possible range of these different colors.

It is correct on a mathematical basis to consider that the color solid comprises infinitely many colors, each color being represented by an infinitely small point in the color space; a point (say x=0.2000,

Fig. 406   A color solid constituted from all the gamuts, each being associated with one particular value of Y between 0.0 and 1.0

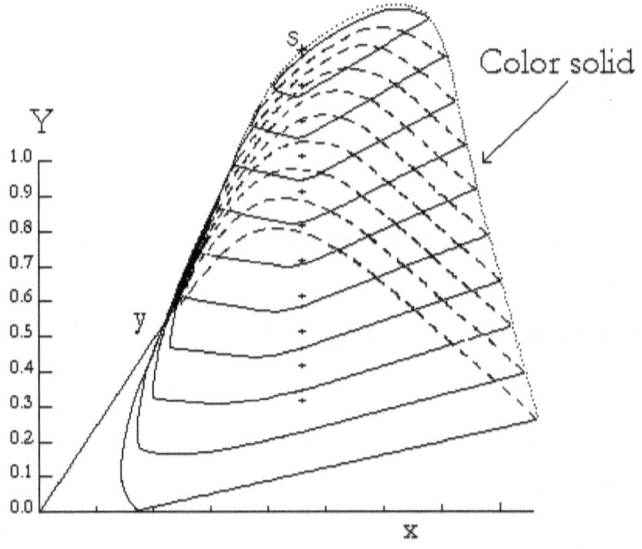

y=0.3000, and Y=0.7) and another point (say x=0.2001, y=0.3001, and Y=0.7) represent two different colors. Actually, such two colors in most situations can only be perceived as the same color rather than two different ones, because human visual system can hardly distinguish between the colors with so small a chromaticity difference.

Imagine that each different color in everyday experiences is represented by a tiny ball of a certain volume in the color space, instead of a geometric point, and the color solid is filled up with these tiny balls. For a broadened color solid, more balls would be needed to fill it up. It means that the color solid of a larger volume would comprise a greater number of different colors. It is in this sense that the following conclusion holds true:

--- The volume of the color solid associated with the spectral power distribution of given light source determines the upper limit for the number of the object colors producible under the given light source with one unit luminous flux, and is proportional to the total number of these different colors.

Now the volume of the color solid is taken to serve as one more measure of the color rendering quality of light, to assess how wide a range of colors can be rendered by a given light source. This measure will be called the Color Rendering Capacity:

--- The Color Rendering Capacity of a light source is a figure representing the volume of the color solid associated with the spectral power distribution of the light source, that determines the upper limit for the range, the "number", of the different colors producible under the light source of given spectral power distribution with one unit luminous flux, and is proportional to the maximum possible range, the "total number", of these colors.

The higher the color rendering capacity, the wider the range of the colors that can be vividly rendered by the given kind of light sources. A light source having high color rendering capacity is capable of rendering a great number of widely different colors and making a colorful scene. Of course, as noted in Section 1.1, two factors are indispensable in determining whether or not a scene looks colorful: the color rendering quality of the illuminating source as well as the color characteristics of the scene itself. A scene composed of gray materials would hardly look colorful, even if illuminated by a source having high color rendering capacity.

Examining the derivation of the color solid, it can be noticed that the volume of the color solid depends solely on the "kind" of the spectral power distribution of light source, independent of the magnitude of spectral power or anything else, and so the color rendering capacity could be regarded as one of the inherent characteristics of light.

4.4 Procedure for Calculating Color Rendering Capacity

The key step in calculation of color rendering capacity is to obtain each maximum chromaticity range, i.e. the gamut associated with one specified value of luminous flux reflectance between 0.0 and 1.0, using the square spectral reflectance curves in conjunction with the spectral power distribution of the given light source, before constructing a color solid with these gamuts and calculating the volume of the color solid in a uniform color space. The calculation will be carried out in the CIE 1976 u*v*L* Uniform Color Space, instead of the previously used x, y, Y color specification system. The transformation involved is as follows:

Suppose X_0, Y_0, Z_0 are the tristimulus values of the light source of given spectral power distribution, and X, Y, Z are the tristimulus values of the reflecting surface illuminated by the given light source.

$u_0' = 4X_0/(X_0 + 15Y_0 + 3Z_0)$, $v_0' = 9Y_0/(X_0 + 15Y_0 + 3Z_0)$

$u' = 4X/(X + 15Y + 3Z)$, $v' = 9Y/(X + 15Y + 3Z)$

$u^* = 13L^*(u' - u_0')$, $v^* = 13L^*(v' - v_0')$

$L^* = 116(Y/Y_0)^{1/3} - 16$

where $Y/Y_0 = \Sigma P(\lambda)R(\lambda)\underline{y}(\lambda)/\Sigma P(\lambda)\underline{y}(\lambda)$, being the luminous flux reflectance of the reflecting surface under the given light source.

The calculation procedure is outlined as follows (see Fig. 407):

Fig. 407   Basic steps of calculating chromaticities having a certain value of luminous flux reflectance

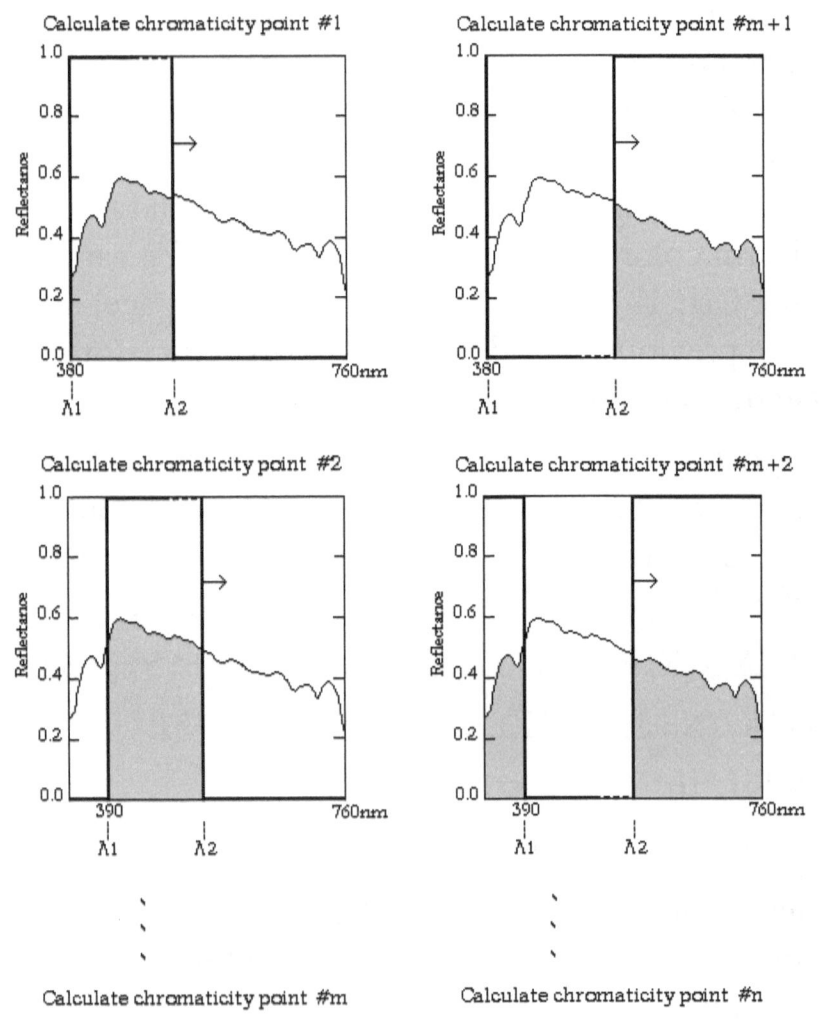

Suppose the spectral power distribution of the given light source is $P(\lambda)$.

$Y_0 = \Sigma P(\lambda)\underline{y}(\lambda)$, ($\lambda$=380nm to 760nm)

$L^* = 0$

DO

$L^* = L^* + 10$

$Y/Y_0 = ((L^*+16)/116)^3$ (Note: Y/Y_0 is the luminous flux reflectance of the reflecting surface under the given light source).

1) Consider the square spectral reflectance curves containing exclusively one reflecting band. Suppose the band begins at $\lambda 1$, and ends at $\lambda 2$.

First, set $\lambda 1$=380nm, $\lambda 2$=380nm, and increase $\lambda 2$ one nm by one nm until:

$\Sigma P(\lambda)\underline{y}(\lambda) = (Y/Y_0)Y_0$, ($\lambda=\lambda 1$ to $\lambda 2$)

Using the square spectral reflectance curve so determined, in conjunction with the given $P(\lambda)$, calculate one set of tristimulus values X,Y,Z, and one chromaticity point (u^*, v^*), associated with the given L^*.

Next, set $\lambda 1$=390nm, $\lambda 2$=390nm, and increase $\lambda 2$ one nm by one nm until:

$\Sigma P(\lambda)\underline{y}(\lambda) = (Y/Y_0)Y_0$, ($\lambda=\lambda 1$ to $\lambda 2$)

Using the square spectral reflectance curve so determined, in conjunction with the given $P(\lambda)$, calculate one more set of tristimulus values X,Y,Z, and one more chromaticity point (u*,v*), associated with the given L*.

. . .

Repeat the calculation until, at a certain λ1, the the equation $\Sigma P(\lambda)\underline{y}(\lambda) = (Y/Y_0)Y_0$, (λ=λ1 to λ2) can no longer be satisfied even if λ2 has been increased to 769nm.

Suppose m chromaticity points so far have been obtained

2) Consider the square spectral reflectance curves containing exclusively one absorbing band. Suppose the band begins at λ1, and ends at λ2.

First, set λ1=380nm, λ2=380nm, and increase λ2 one nm by one nm until:

$\Sigma P(\lambda)\underline{y}(\lambda) = (1-(Y/Y_0))Y_0$, (λ=λ1 to λ2)

Using the square spectral reflectance curve so determined, in conjunction with the given P(λ), calculate another set of tristimulus values X,Y,Z, and another chromaticity point (u*,v*) numbered m+1, associated with the given L*.

Next, set λ1=390nm, λ2=390nm, and increase λ2 one nm by one nm until:

$\Sigma P(\lambda)\underline{y}(\lambda) = (1-(Y/Y_0))Y_0$, (λ=λ1 to λ2)

Using the square spectral reflectance curve so determined, in conjunction with the given P(λ), calculate still another set of tristimulus values

X,Y,Z, and still another chromaticity point (u^*,v^*) numbered m+2, associated with the given L*.

. . .

Repeat the calculation until, at a certain $\lambda 1$, the equation $\Sigma P(\lambda)\underline{y}(\lambda) = (1-(Y/Y_0))Y_0$, $(\lambda=\lambda 1$ to $\lambda 2)$ can no longer be satisfied even if $\lambda 2$ has been increased to 769nm.

Suppose altogether n chromaticity points so far have been obtained. From these n chromaticity points, a polygon can be constructed on a (u^*, v^*) right-angled coordinate plane. Calculate the area of the polygon, which is the maximum chromaticity range or gamut associated with the given value of L* :

gamut(L*) = the area of the polygon associated with the given value of L*.

LOOP UNTIL L*=90

The volume of the color solid associated with the spectral power distribution $P(\lambda)$ of the given light source = Σ gamut(L*), (L*=10, 20, …, and 90).

The Color Rendering Capacity of the given light source = the volume of the color solid associated with the given $P(\lambda)$ divided by the volume of the color solid associated with equal-energy spectrum.

A software CRC99 for calculating the color rendering capacity of any light source of given spectral composition has been available. A simplified version of CRC99 (crc1.zip, 1.51MB) can be downloaded from the following address:

http://host-a.net/colorrendering/crc1.zip

http://color.rendering.tripod.com

http://knol.google.com/k/xu-he/color-rendering-capacity

How to use software CRC99:

1) Download the file crc1.zip to drive C:\, and unzip the file to folder C:\crc1\;

1) Run the program setup.exe in that folder to install CRC99;

2) To run CRC99: click Start – Programs – Crc99 – Crc99. That is all!

4.5 Examples of Color Rendering Capacity Calculation

Shown in Fig. 408 through Fig. 415 are reports of CRC99 calculating the color rendering capacity for several spectral power distributions, including equal-energy spectrum, CIE illuminants A and D65, high pressure sodium, fluorescent lamps F2, F7, and F11, and V-shaped spectrum.

Fig. 408 Result of color rendering capacity calculatiom for equal-energy spectrum

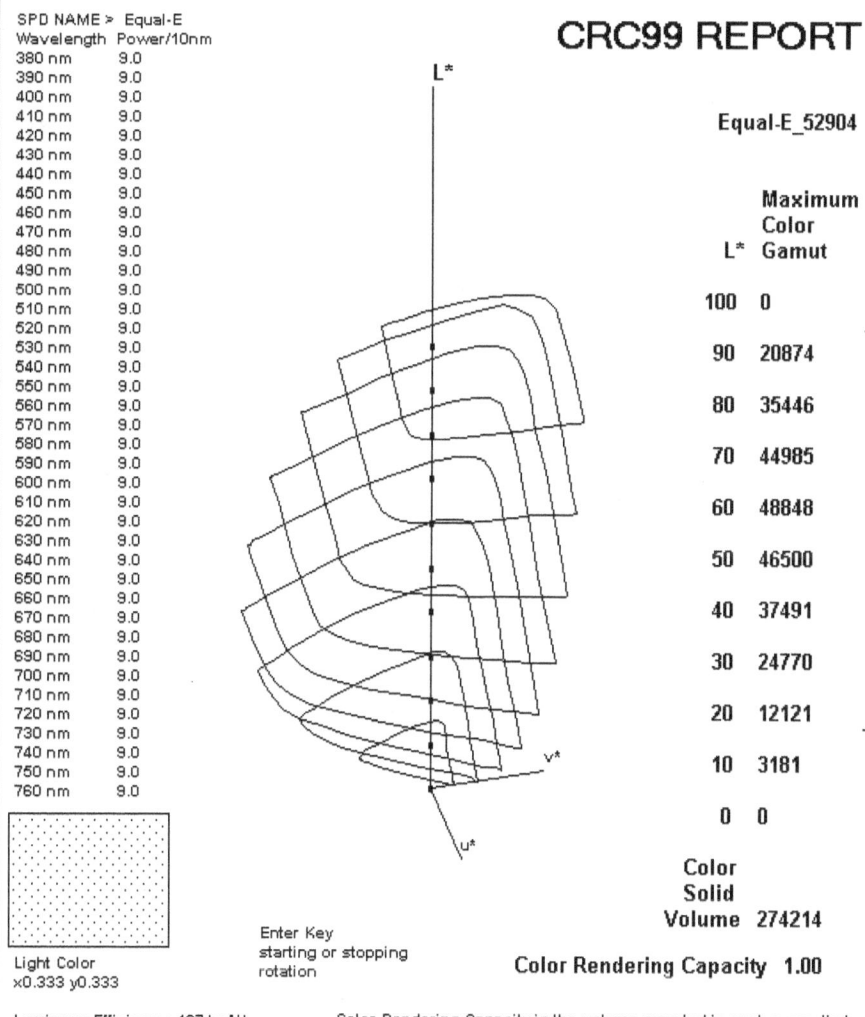

SPD NAME > Equal-E
Wavelength Power/10nm

Wavelength	Power/10nm
380 nm	9.0
390 nm	9.0
400 nm	9.0
410 nm	9.0
420 nm	9.0
430 nm	9.0
440 nm	9.0
450 nm	9.0
460 nm	9.0
470 nm	9.0
480 nm	9.0
490 nm	9.0
500 nm	9.0
510 nm	9.0
520 nm	9.0
530 nm	9.0
540 nm	9.0
550 nm	9.0
560 nm	9.0
570 nm	9.0
580 nm	9.0
590 nm	9.0
600 nm	9.0
610 nm	9.0
620 nm	9.0
630 nm	9.0
640 nm	9.0
650 nm	9.0
660 nm	9.0
670 nm	9.0
680 nm	9.0
690 nm	9.0
700 nm	9.0
710 nm	9.0
720 nm	9.0
730 nm	9.0
740 nm	9.0
750 nm	9.0
760 nm	9.0

Light Color
x0.333 y0.333

Luminous Efficiency 187 lm/W

CRC99 REPORT

Equal-E_52904

Maximum
Color
L* Gamut

L*	Maximum Color Gamut
100	0
90	20874
80	35446
70	44985
60	48848
50	46500
40	37491
30	24770
20	12121
10	3181
0	0

Color
Solid
Volume 274214

Color Rendering Capacity 1.00

Enter Key
starting or stopping
rotation

Color Rendering Capacity is the volume reported in such a way that
the volume with equi-energy spectrum can be regarded as 1.00

Fig. 409 Result of color rendering capacity calculatiom for CIE illuminant A

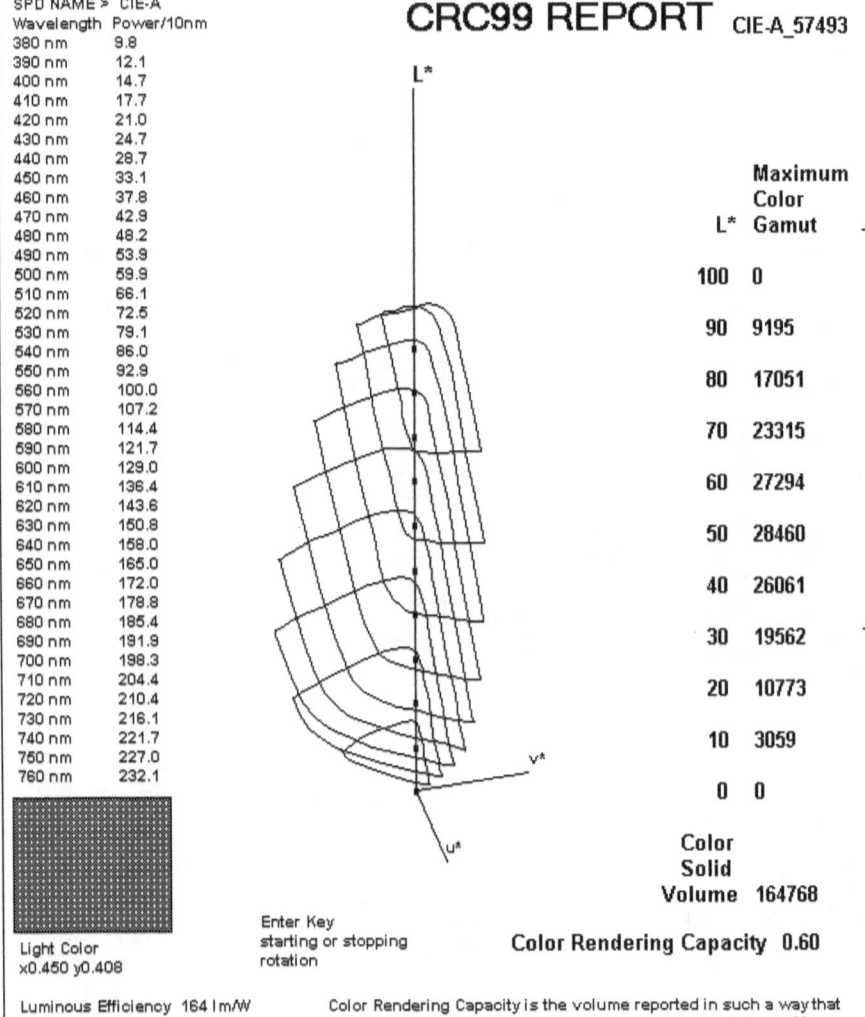

SPD NAME > CIE-A
Wavelength Power/10nm
380 nm   9.8
390 nm   12.1
400 nm   14.7
410 nm   17.7
420 nm   21.0
430 nm   24.7
440 nm   28.7
450 nm   33.1
460 nm   37.8
470 nm   42.9
480 nm   48.2
490 nm   53.9
500 nm   59.9
510 nm   66.1
520 nm   72.5
530 nm   79.1
540 nm   86.0
550 nm   92.9
560 nm   100.0
570 nm   107.2
580 nm   114.4
590 nm   121.7
600 nm   129.0
610 nm   136.4
620 nm   143.6
630 nm   150.8
640 nm   158.0
650 nm   165.0
660 nm   172.0
670 nm   178.8
680 nm   185.4
690 nm   191.9
700 nm   198.3
710 nm   204.4
720 nm   210.4
730 nm   216.1
740 nm   221.7
750 nm   227.0
760 nm   232.1

Light Color
x0.450 y0.408

Luminous Efficiency 164 lm/W

CRC99 REPORT CIE-A_57493

	Maximum Color
L*	Gamut
100	0
90	9195
80	17051
70	23315
60	27294
50	28460
40	26061
30	19562
20	10773
10	3059
0	0

Color Solid Volume 164768

Enter Key
starting or stopping
rotation

Color Rendering Capacity 0.60

Color Rendering Capacity is the volume reported in such a way that the volume with equi-energy spectrum can be regarded as 1.00

Fig. 410   Result of color rendering capacity calculatiom
for CIE illuminant D65

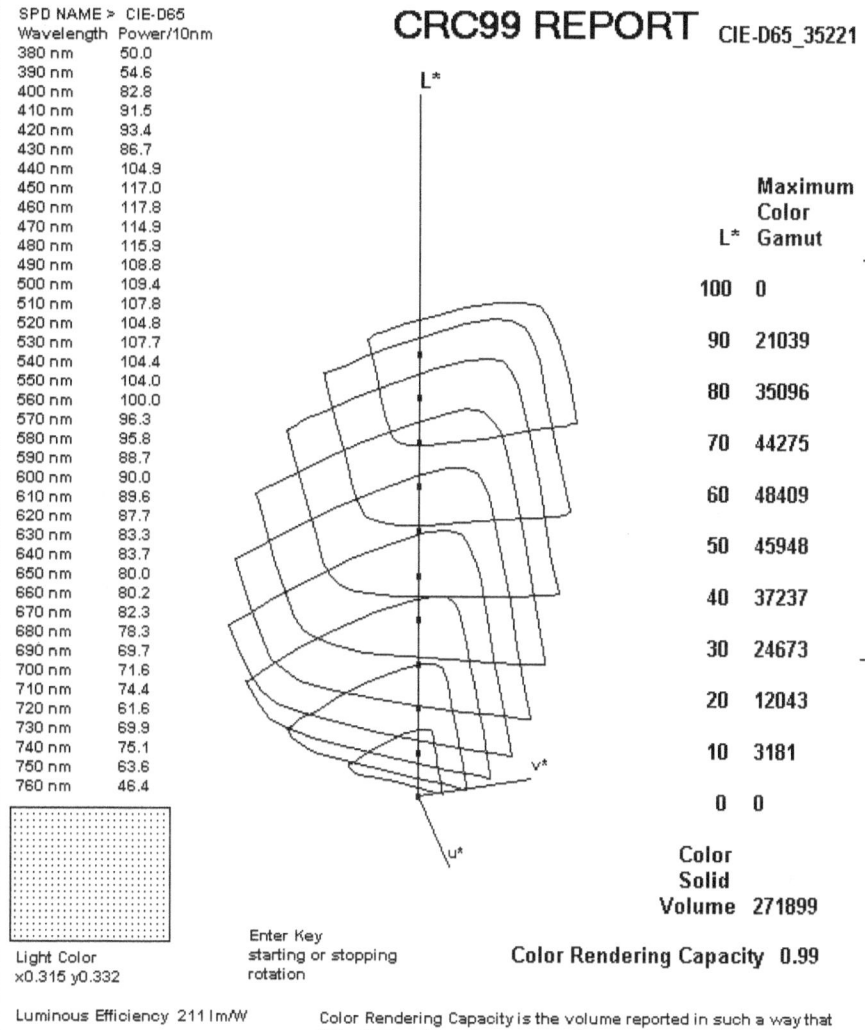

SPD NAME > CIE-D65

Wavelength	Power/10nm
380 nm	50.0
390 nm	54.6
400 nm	82.8
410 nm	91.5
420 nm	93.4
430 nm	86.7
440 nm	104.9
450 nm	117.0
460 nm	117.8
470 nm	114.9
480 nm	115.9
490 nm	108.8
500 nm	109.4
510 nm	107.8
520 nm	104.8
530 nm	107.7
540 nm	104.4
550 nm	104.0
560 nm	100.0
570 nm	96.3
580 nm	95.8
590 nm	88.7
600 nm	90.0
610 nm	89.6
620 nm	87.7
630 nm	83.3
640 nm	83.7
650 nm	80.0
660 nm	80.2
670 nm	82.3
680 nm	78.3
690 nm	69.7
700 nm	71.6
710 nm	74.4
720 nm	61.6
730 nm	69.9
740 nm	75.1
750 nm	63.6
760 nm	46.4

CRC99 REPORT   CIE-D65_35221

L*

Maximum
Color
L*   Gamut

L*	Gamut
100	0
90	21039
80	35096
70	44275
60	48409
50	45948
40	37237
30	24673
20	12043
10	3181
0	0

Color
Solid
Volume 271899

Light Color
x0.315 y0.332

Enter Key
starting or stopping
rotation

Color Rendering Capacity 0.99

Luminous Efficiency 211 lm/W

Color Rendering Capacity is the volume reported in such a way that
the volume with equi-energy spectrum can be regarded as 1.00

Fig. 411 Result of color rendering capacity calculatiom for High pressure sodium

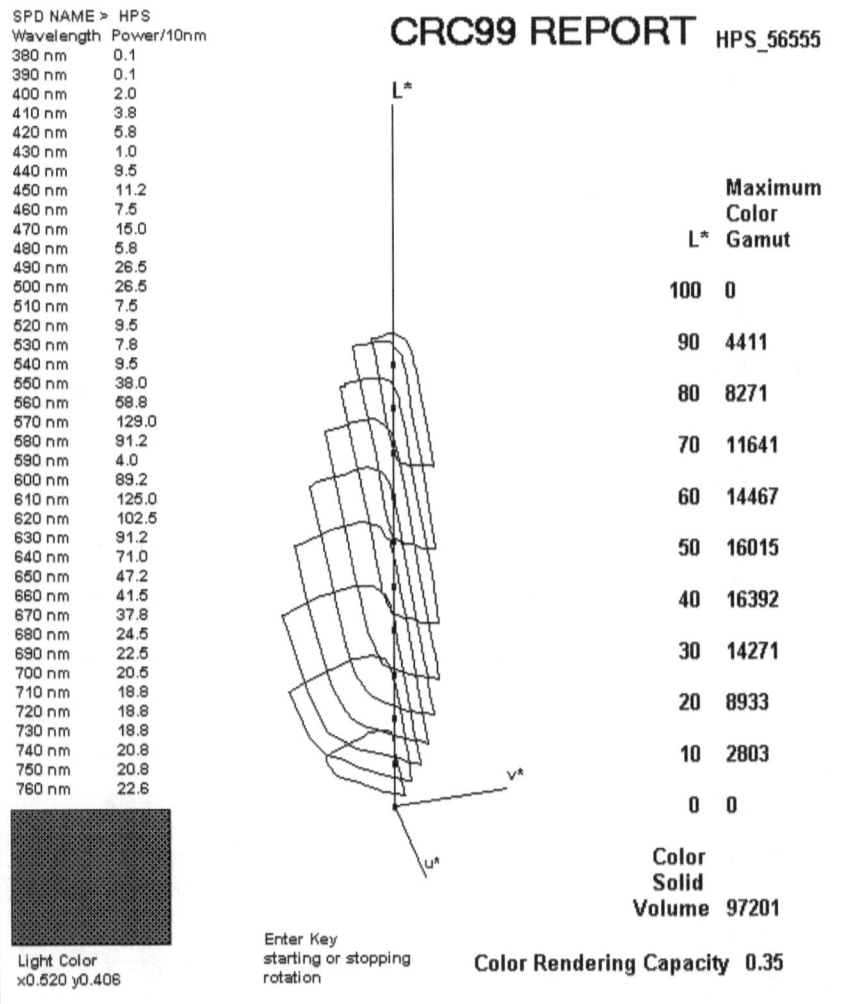

SPD NAME > HPS

Wavelength	Power/10nm
380 nm	0.1
390 nm	0.1
400 nm	2.0
410 nm	3.8
420 nm	5.8
430 nm	1.0
440 nm	9.5
450 nm	11.2
460 nm	7.5
470 nm	15.0
480 nm	5.8
490 nm	26.5
500 nm	26.5
510 nm	7.5
520 nm	9.5
530 nm	7.8
540 nm	9.5
550 nm	38.0
560 nm	58.8
570 nm	129.0
580 nm	91.2
590 nm	4.0
600 nm	89.2
610 nm	125.0
620 nm	102.5
630 nm	91.2
640 nm	71.0
650 nm	47.2
660 nm	41.5
670 nm	37.8
680 nm	24.5
690 nm	22.5
700 nm	20.5
710 nm	18.8
720 nm	18.8
730 nm	18.8
740 nm	20.8
750 nm	20.8
760 nm	22.6

Light Color
x0.520 y0.406

Luminous Efficiency 281 lm/W

CRC99 REPORT HPS_56555

L^*

	Maximum Color
L^*	Gamut
100	0
90	4411
80	8271
70	11641
60	14467
50	16015
40	16392
30	14271
20	8933
10	2803
0	0

v^*

u^*

Color
Solid
Volume 97201

Enter Key
starting or stopping
rotation

Color Rendering Capacity 0.35

Color Rendering Capacity is the volume reported in such a way that
the volume with equi-energy spectrum can be regarded as 1.00

Fig. 412   Result of color rendering capacity calculatiom for Normal fluorescent (F2)

SPD NAME > F2

Wavelength	Power/10nm
380 nm	2.7
390 nm	4.1
400 nm	19.1
410 nm	7.6
420 nm	8.8
430 nm	40.0
440 nm	18.1
450 nm	13.6
460 nm	14.6
470 nm	15.2
480 nm	15.3
490 nm	15.1
500 nm	14.4
510 nm	14.1
520 nm	14.6
530 nm	16.9
540 nm	34.9
550 nm	31.2
560 nm	33.7
570 nm	40.1
580 nm	42.1
590 nm	36.4
600 nm	31.8
610 nm	26.2
620 nm	20.6
630 nm	15.7
640 nm	11.7
650 nm	8.7
660 nm	6.4
670 nm	4.7
680 nm	3.5
690 nm	2.8
700 nm	2.1
710 nm	1.6
720 nm	1.3
730 nm	1.1
740 nm	1.0
750 nm	0.9
760 nm	0.9

Light Color
x0.376 y0.374

Luminous Efficiency 334 lm/W

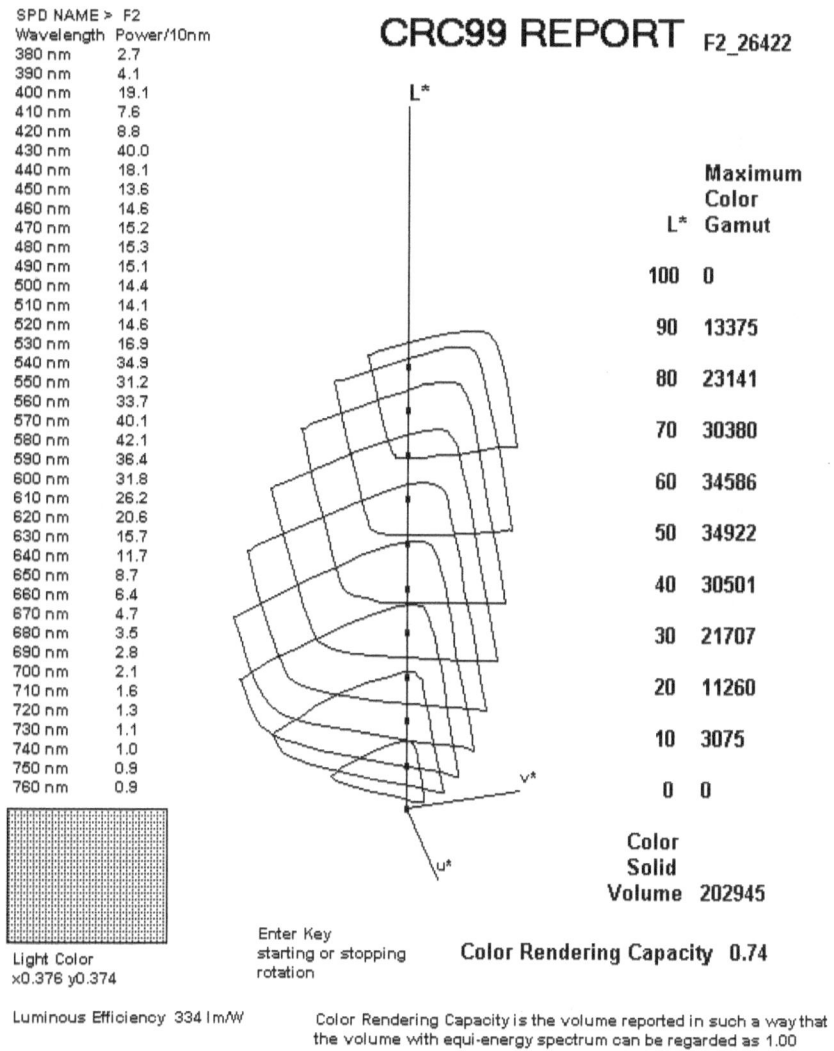

CRC99 REPORT   F2_26422

L*

Maximum
Color
L* Gamut

L*	Gamut
100	0
90	13375
80	23141
70	30380
60	34586
50	34922
40	30501
30	21707
20	11260
10	3075
0	0

Color
Solid
Volume 202945

Enter Key
starting or stopping
rotation

Color Rendering Capacity 0.74

Color Rendering Capacity is the volume reported in such a way that
the volume with equi-energy spectrum can be regarded as 1.00

Fig. 413 Result of color rendering capacity calculatiom for De luxe fluorescent (F7)

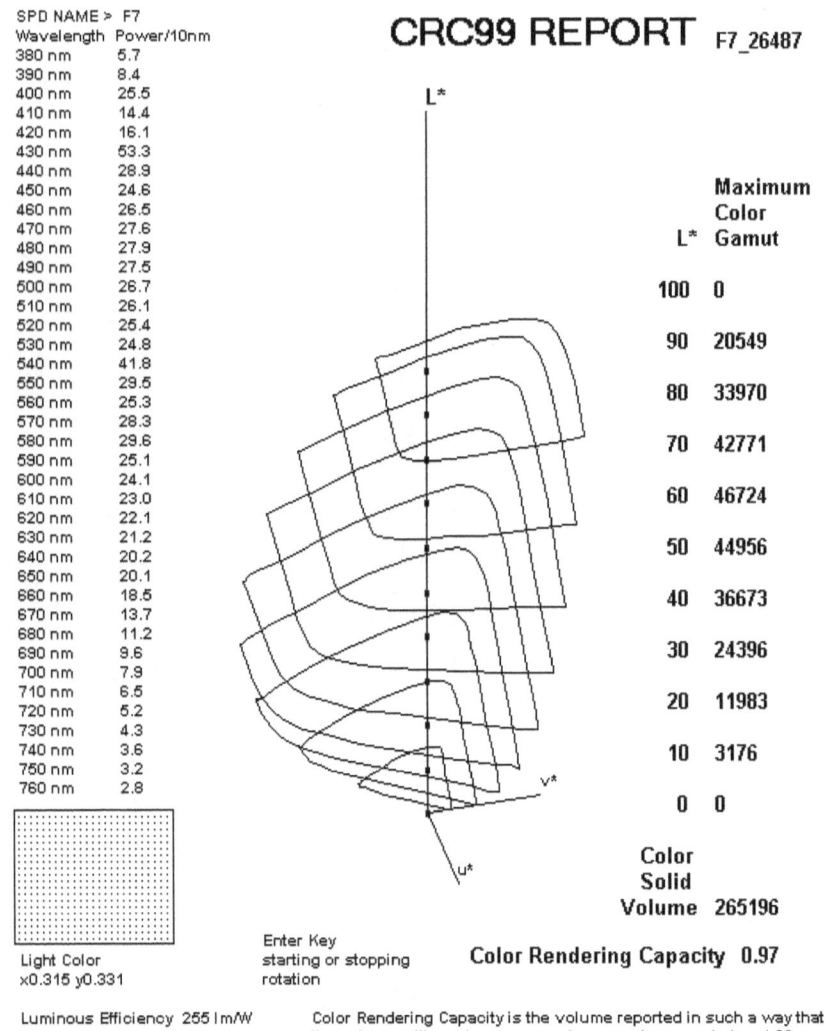

SPD NAME > F7
Wavelength Power/10nm

Wavelength	Power/10nm
380 nm	5.7
390 nm	8.4
400 nm	25.5
410 nm	14.4
420 nm	16.1
430 nm	53.3
440 nm	28.9
450 nm	24.6
460 nm	26.5
470 nm	27.6
480 nm	27.9
490 nm	27.5
500 nm	26.7
510 nm	26.1
520 nm	25.4
530 nm	24.8
540 nm	41.8
550 nm	29.5
560 nm	25.3
570 nm	28.3
580 nm	29.6
590 nm	25.1
600 nm	24.1
610 nm	23.0
620 nm	22.1
630 nm	21.2
640 nm	20.2
650 nm	20.1
660 nm	18.5
670 nm	13.7
680 nm	11.2
690 nm	9.6
700 nm	7.9
710 nm	6.5
720 nm	5.2
730 nm	4.3
740 nm	3.6
750 nm	3.2
760 nm	2.8

Light Color
x0.315 y0.331

Luminous Efficiency 255 lm/W

Enter Key
starting or stopping
rotation

CRC99 REPORT F7_26487

L*

**Maximum
Color
L* Gamut**

L*	Gamut
100	0
90	20549
80	33970
70	42771
60	46724
50	44956
40	36673
30	24396
20	11983
10	3176
0	0

v*

u*

**Color
Solid
Volume 265196**

Color Rendering Capacity 0.97

Color Rendering Capacity is the volume reported in such a way that
the volume with equi-energy spectrum can be regarded as 1.00

Fig. 414 Result of color rendering capacity calculatiom for Tri-band fluorescent (F11)

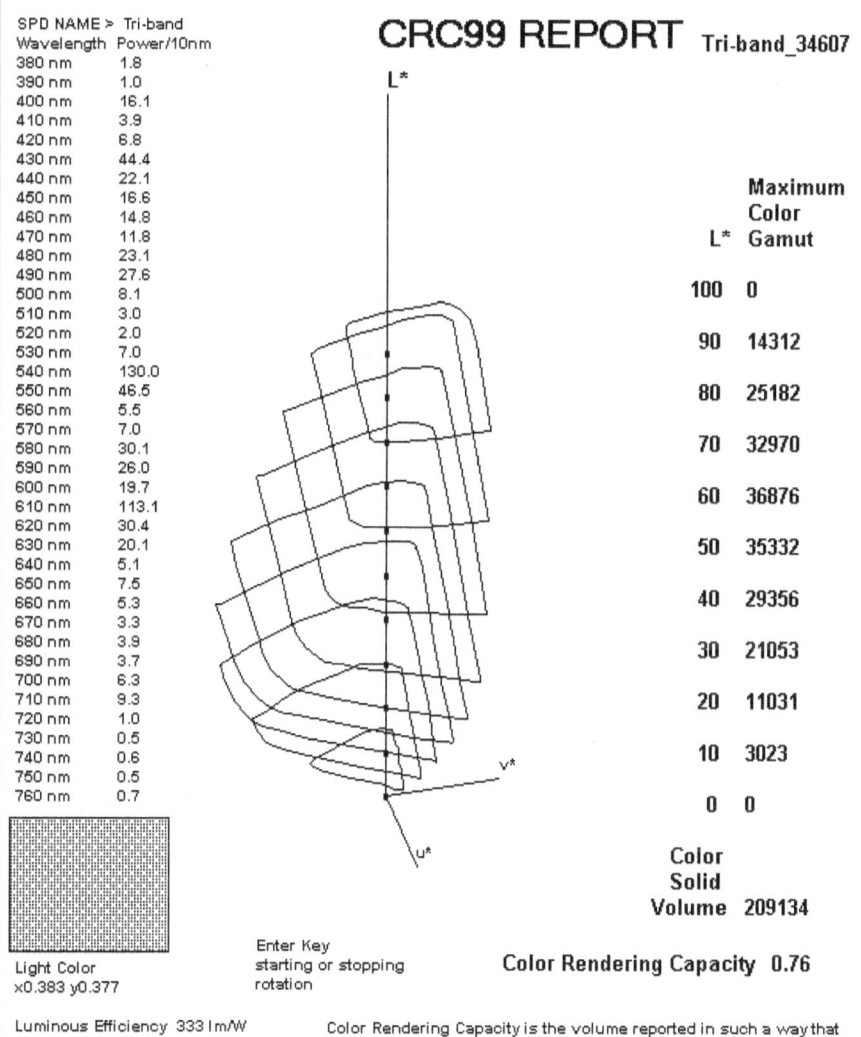

SPD NAME > Tri-band	
Wavelength	Power/10nm
380 nm	1.8
390 nm	1.0
400 nm	16.1
410 nm	3.9
420 nm	6.8
430 nm	44.4
440 nm	22.1
450 nm	16.6
460 nm	14.8
470 nm	11.8
480 nm	23.1
490 nm	27.6
500 nm	8.1
510 nm	3.0
520 nm	2.0
530 nm	7.0
540 nm	130.0
550 nm	46.5
560 nm	5.5
570 nm	7.0
580 nm	30.1
590 nm	26.0
600 nm	19.7
610 nm	113.1
620 nm	30.4
630 nm	20.1
640 nm	5.1
650 nm	7.5
660 nm	5.3
670 nm	3.3
680 nm	3.9
690 nm	3.7
700 nm	6.3
710 nm	9.3
720 nm	1.0
730 nm	0.5
740 nm	0.6
750 nm	0.5
760 nm	0.7

Light Color
x0.383 y0.377

Luminous Efficiency 333 lm/W

Enter Key
starting or stopping
rotation

CRC99 REPORT Tri-band_34607

L*

	Maximum Color
L*	Gamut
100	0
90	14312
80	25182
70	32970
60	36876
50	35332
40	29356
30	21053
20	11031
10	3023
0	0

v*

u*

Color Solid Volume 209134

Color Rendering Capacity 0.76

Color Rendering Capacity is the volume reported in such a way that
the volume with equi-energy spectrum can be regarded as 1.00

Fig. 415 Result of color rendering capacity calculatiom
for V-shaped spectrum

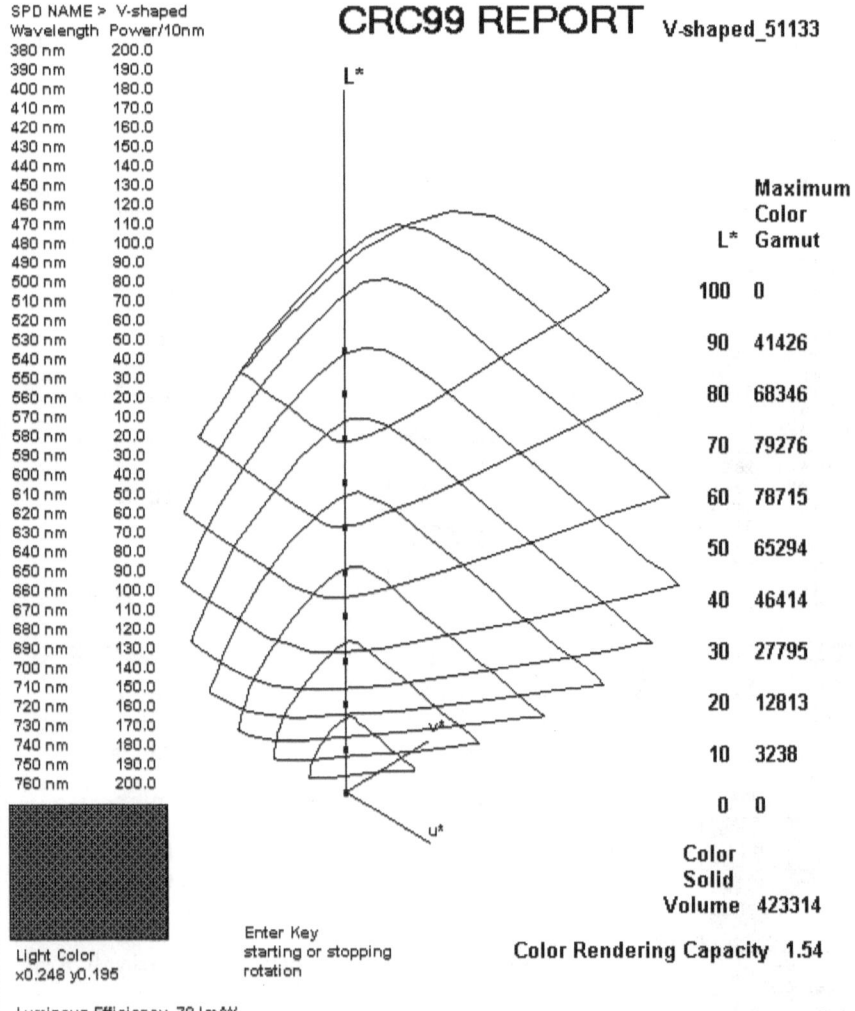

SPD NAME > V-shaped
Wavelength Power/10nm
380 nm   200.0
390 nm   190.0
400 nm   180.0
410 nm   170.0
420 nm   160.0
430 nm   150.0
440 nm   140.0
450 nm   130.0
460 nm   120.0
470 nm   110.0
480 nm   100.0
490 nm   90.0
500 nm   80.0
510 nm   70.0
520 nm   60.0
530 nm   50.0
540 nm   40.0
550 nm   30.0
560 nm   20.0
570 nm   10.0
580 nm   20.0
590 nm   30.0
600 nm   40.0
610 nm   50.0
620 nm   60.0
630 nm   70.0
640 nm   80.0
650 nm   90.0
660 nm   100.0
670 nm   110.0
680 nm   120.0
690 nm   130.0
700 nm   140.0
710 nm   150.0
720 nm   160.0
730 nm   170.0
740 nm   180.0
750 nm   190.0
760 nm   200.0

Light Color
x0.248 y0.195

Luminous Efficiency 79 lm/W

CRC99 REPORT V-shaped_51133

L*	Maximum Color Gamut
100	0
90	41426
80	68346
70	79276
60	78715
50	65294
40	46414
30	27795
20	12813
10	3238
0	0

Color Solid Volume 423314

Enter Key
starting or stopping
rotation

Color Rendering Capacity 1.54

Color Rendering Capacity is the volume reported in such a way that
the volume with equi-energy spectrum can be regarded as 1.00

4.6 Supportive Information about Color Rendering Capacity

Since high color rendering capacity always reflects broadened color solid, the interior surfaces under illumination of high color rendering capacity may be able to exhibit large chroma C^*uv, i.e. large $((u^*)^2+(v^*)^2)^{1/2}$, on average.

A computer-simulated experiment was conducted to test a similar prediction, see Reference 11. In the experiment, eight spectral power distributions of different color rendering capacity are involved, including fluorescent lamps F2, F6, F7, F8, F10, and F11, CIE illuminant A, and high pressure sodium. Ten rooms are used, each being composed of 300 reflecting surfaces, represented by 300 randomly generated spectral reflectance curves. The experiment proceeds like this: illuminate a room with each of eight light sources in turn while calculating the C^*uv for each of 300 surfaces before calculating one average C^*uv over the 300 surfaces making up the room for each light source. Repeat the process for each of ten rooms at a fixed illuminance. The collected data are shown in Fig. 416, and will be used in the following study.

Having no knowledge about the population normality of the data, the following study chooses to perform a k-sample nonparametric statistical test, the Page trend test, to see whether the average C^*uv of interior surfaces increases with the color rendering

Fig. 416 Information about a computer-simulated experiment

Spectral power distribution (SPD) having different color rendering capacity (CRC)

SPD	HPS	A	F6	F2	F11	F10	F8	F7
CRC	0.35	0.60	0.68	0.74	0.76	0.83	0.90	0.97

One of 300 randomly generated spectral reflectance curves making up a room

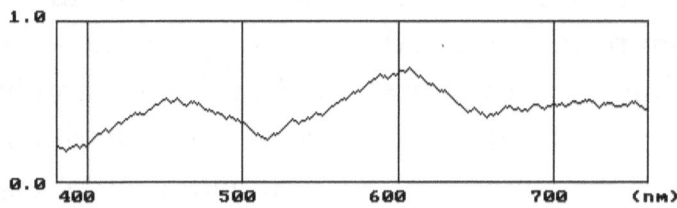

The average C*uv of interior surfaces under different light sources

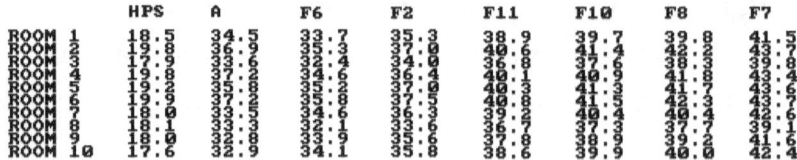

Page test statistic L = 2031
(At a significance level of 0.001 for k=8 and N=10, the critical value of L is 1773)

capacity of illuminating light sources, the null hypothesis being that there is no such trend across the eight groups of data.

The result shows that the observed Page statistic L is large enough for the null hypothesis to be rejected at a significance level of 0.001 in favor of the alternative hypothesis, that is, large average C^*uv goes with higher color rendering capacity. In a sense, this result supports the statement made in Section 4.3: a light source of high color rendering capacity is capable of making a colorful scene.

4.7 Color Rendering Capacity Highlights Effectiveness of Color Rendering

Both the CIE color rendering index (CRI) and the color rendering capacity (CRC) are measures of the color rendering quality of light, each being focuses on one different aspect of the subject. What the CRI concerns is whether a given light source can render colors in the same way as an incandescent reference illuminant does, while the CRC is concerned with whether a given light source is capable of rendering a great number of widely different colors. In terms of communication theory, the difference between CRI and CRC may be described as follows: the CRI is concerning the "accuracy" of color rendering, while the CRC is concerning the "effectiveness" of color rendering.

The effectiveness of color rendering refers to the ability of light to render surfaces of different spectral reflectance characteristics in widely different colors. Here the phrase "different colors" deserves particular attention.

Let two different colors be represented by two points some distance apart in a uniform color space, and the extent to which the two colors are different from each other be represented by the distance between the two points. Theoretically, any two colors could be called "different" so long as the two points representing them are not merging into a single one. As a matter of fact, long before the distance between the two points decreases to a certain value, the two colors have not been visually distinguishable between each other.

Suppose there is a collection of reflecting surfaces, each being characterized by one unique spectral reflectance curve. Consider the set R of all spectral reflectance curves possible in Physics, R={r1, r2, ...}. This is an infinite set comprising infinitely many elements.

Under illumination of a certain spectral power distribution $P(\lambda)$, the infinite set R of all spectral reflectance curves is mapped one-to-one onto an infinite set C of colors, C={c1, c2, ...}, represented by a color solid of finite volume in a color space.

Under a $P(\lambda)$ of high color rendering capacity, the infinite set C of colors is represented by a broadened color solid with a larger volume. This means that surfaces of different spectral reflectance characteristics will correspond to points a bigger distance apart in the color space, and these points represent widely different colors.

Under a P(λ) of low color rendering capacity, the infinite set C of colors is represented by a narrowed color solid. Imagine that the infinite set R of all spectral reflectance curves, or the collection of reflecting surfaces characterized by these different spectral reflectance curves, is being squeezed into a narrower color solid. In consequence, more surfaces of different spectral reflectance characteristics will be rendered in colors so resembling each other (the points representing these colors are spaced so close to each other in the color space) that these surfaces of different spectral reflectance characteristics look almost identical in color appearance, even if those colors remain different in theory.

Where the effectiveness of color rendering is concerned, a light source can be rated as a good one only if it is able to render surfaces of different spectral reflectance characteristics in widely different colors, to make those points representing these colors be away from each other as far as possible. Such result can be achieved only when there is a color solid of large volume, and this means that the light source, or the spectral power distribution, should be of high color rendering capacity.

Appendix 1

Data of spectral power distributions P(λ) for CIE illuminants A and D65, High pressure sodium, and Tri-band fluorescent (F11) at 10nm wavelength intervals

Wavelength (nm)	A	D65	HPS	F11
380	9.8	50.0	0.0	1.8
390	12.1	54.6	0.0	1.0
400	14.7	82.8	2.0	16.1
410	17.7	91.5	3.8	3.9
420	21.0	93.4	5.8	6.8
430	24.7	86.7	1.0	44.4
440	28.7	104.9	9.5	22.1
450	33.1	117.0	11.2	16.6
460	37.8	117.8	7.5	14.8
470	42.9	114.9	15.0	11.8
480	48.2	115.9	5.8	23.1
490	53.9	108.8	26.5	27.6
500	59.9	109.4	26.5	8.1
510	66.1	107.8	7.5	3.0
520	72.5	104.8	9.5	2.0
530	79.1	107.7	7.8	7.0
540	86.0	104.4	9.5	130.0
550	92.9	104.0	38.0	46.5
560	100.0	100.0	58.8	5.5
570	107.2	96.3	129.0	7.0
580	114.4	95.8	91.2	30.1
590	121.7	88.7	4.0	26.0
600	129.0	90.0	89.2	19.7
610	136.4	89.6	125.0	113.1
620	143.6	87.7	102.5	30.4
630	150.8	83.3	91.2	20.1
640	158.0	83.7	71.0	5.1
650	165.0	80.0	47.2	7.5
660	172.0	80.2	41.5	5.3
670	178.8	82.3	37.8	3.3
680	185.4	78.3	24.5	3.9
690	191.9	69.7	22.5	3.7
700	198.3	71.6	20.5	6.3
710	204.4	74.4	18.8	9.3
720	210.4	61.6	18.8	1.0
730	216.1	69.9	18.8	0.5
740	221.7	75.1	20.8	0.6
750	227.0	63.6	20.8	0.5
760	232.1	46.4	22.6	0.7

Appendix 2

Data of CIE 1931 standard observer color-matching functions, i.e. tristimulus values of the spectrum colors $\bar{x}(\lambda)$, $\bar{y}(\lambda)$, and $\bar{z}(\lambda)$ at 1nm wavelength intervals

Wavelength (nm)	$\bar{x}(\lambda)$	$\bar{y}(\lambda)$	$\bar{z}(\lambda)$
380	0.0013680	0.0000390	0.0004500
381	0.0015020	0.0000428	0.0070832
382	0.0016423	0.0000469	0.0077455
383	0.0018024	0.0000516	0.0085012
384	0.0019958	0.0000572	0.0094145
385	0.0022360	0.0000640	0.0105500
386	0.0025354	0.0000723	0.0119658
387	0.0028926	0.0000822	0.0136559
388	0.0033008	0.0000935	0.0155880
389	0.0037532	0.0001061	0.0177302
390	0.0042430	0.0001200	0.0200500
391	0.0047624	0.0001350	0.0225114
392	0.0053300	0.0001515	0.0252029
393	0.0059787	0.0001702	0.0282797
394	0.0067411	0.0001918	0.0318970
395	0.0076500	0.0002170	0.0362100
396	0.0087514	0.0002469	0.0414377
397	0.0100289	0.0002812	0.0475037
398	0.0114217	0.0003185	0.0541199
399	0.0128690	0.0003573	0.0609980
400	0.0143100	0.0003960	0.0678500
401	0.0157044	0.0004337	0.0744863
402	0.0171474	0.0004730	0.0813616
403	0.0187812	0.0005179	0.0891536
404	0.0207480	0.0005722	0.0985405
405	0.0231900	0.0006400	0.1102000
406	0.0262074	0.0007246	0.1246133
407	0.0297825	0.0008255	0.1417017
408	0.0338809	0.0009412	0.1613035
409	0.0384682	0.0010699	0.1832563
410	0.0435100	0.0012100	0.2074000
411	0.0489956	0.0013621	0.2336921
412	0.0550226	0.0015308	0.2626114
413	0.0617188	0.0017204	0.2947746
414	0.0692120	0.0019353	0.3307985
415	0.0776300	0.0021800	0.3713000
416	0.0869581	0.0024548	0.4162091
417	0.0971767	0.0027640	0.4654642
418	0.1084063	0.0031178	0.5196948
419	0.1207672	0.0035264	0.5795303

(to be continued)

(continued)

Wavelength (nm)	$\overline{x}(\lambda)$	$\overline{y}(\lambda)$	$\overline{z}(\lambda)$
420	0.1343800	0.0040000	0.6456000
421	0.1493582	0.0045462	0.7184838
422	0.1653957	0.0051593	0.7967133
423	0.1819831	0.0058293	0.8778459
424	0.1986110	0.0065462	0.9594390
425	0.2147700	0.0073000	1.0390501
426	0.2301868	0.0080865	1.1153673
427	0.2448797	0.0089087	1.1884971
428	0.2587773	0.0097677	1.2581233
429	0.2718079	0.0106644	1.3239296
430	0.2839000	0.0116000	1.3856000
431	0.2949438	0.0125732	1.4426352
432	0.3048965	0.0135827	1.4948035
433	0.3137873	0.0146297	1.5421903
434	0.3216454	0.0157151	1.5848807
435	0.3285000	0.0168400	1.6229600
436	0.3343513	0.0180074	1.6564048
437	0.3392101	0.0192145	1.6852959
438	0.3431213	0.0204539	1.7098745
439	0.3461296	0.0217182	1.7303821
440	0.3482800	0.0230000	1.7470600
441	0.3495999	0.0242946	1.7600446
442	0.3501474	0.0256102	1.7696233
443	0.3500130	0.0269586	1.7762637
444	0.3492870	0.0283512	1.7804334
445	0.3480600	0.0298000	1.7826000
446	0.3463733	0.0313108	1.7829682
447	0.3442624	0.0328837	1.7816998
448	0.3418088	0.0345211	1.7791982
449	0.3390941	0.0362257	1.7758671
450	0.3362000	0.0380000	1.7721100
451	0.3331977	0.0398467	1.7682589
452	0.3300411	0.0417680	1.7640390
453	0.3266357	0.0437660	1.7589438
454	0.3228868	0.0458427	1.7524663
455	0.3187000	0.0480000	1.7441000
456	0.3140251	0.0502437	1.7335595
457	0.3088840	0.0525730	1.7208581
458	0.3032904	0.0549806	1.7059369
459	0.2972579	0.0574587	1.6887372
460	0.2908000	0.0600000	1.6692000
461	0.2839701	0.0626020	1.6475287
462	0.2767214	0.0652775	1.6234127
463	0.2689178	0.0680121	1.5960223
464	0.2604227	0.0709111	1.5645280
465	0.2511000	0.0739000	1.5281000
466	0.2408475	0.0770160	1.4861114
467	0.2298512	0.0802664	1.4395215
468	0.2184072	0.0836668	1.3898799
469	0.2068115	0.0872328	1.3387362

(to be continued)

(continued)

Wavelength (nm)	$\overline{x}(\lambda)$	$\overline{y}(\lambda)$	$\overline{z}(\lambda)$
470	0.1953600	0.0909800	1.2876400
471	0.1842136	0.0949176	1.2374223
472	0.1733273	0.0990458	1.1878243
473	0.1626881	0.1033674	1.1387611
474	0.1522833	0.1078846	1.0901480
475	0.1421000	0.1126000	1.0419000
476	0.1321786	0.1175320	0.9941976
477	0.1225696	0.1226744	0.9473473
478	0.1132752	0.1279928	0.9014531
479	0.1042979	0.1334528	0.8566193
480	0.0956400	0.1390200	0.8129501
481	0.0872996	0.1446764	0.7705173
482	0.0793080	0.1504693	0.7294448
483	0.0717178	0.1564619	0.6899136
484	0.0645810	0.1627177	0.6521049
485	0.0579500	0.1693000	0.6162000
486	0.0518621	0.1762431	0.5823286
487	0.0462815	0.1835581	0.5504162
488	0.0411509	0.1912735	0.5203376
489	0.0364128	0.1994180	0.4919673
490	0.0320100	0.2080200	0.4651800
491	0.0279172	0.2171199	0.4399246
492	0.0241444	0.2267345	0.4161836
493	0.0206870	0.2368571	0.3938822
494	0.0175404	0.2474812	0.3729459
495	0.0147000	0.2586000	0.3533000
496	0.0121618	0.2701849	0.3348578
497	0.0099200	0.2822939	0.3175521
498	0.0079672	0.2950505	0.3013375
499	0.0062963	0.3085780	0.2861686
500	0.0049000	0.3230000	0.2720000
501	0.0037772	0.3384021	0.2588171
502	0.0029453	0.3546858	0.2464838
503	0.0024249	0.3716986	0.2347718
504	0.0022363	0.3892875	0.2234533
505	0.0024000	0.4073000	0.2123000
506	0.0029255	0.4256299	0.2011692
507	0.0038366	0.4443096	0.1901196
508	0.0051748	0.4633944	0.1792254
509	0.0069821	0.4829395	0.1685608
510	0.0093000	0.5030000	0.1582000
511	0.0121495	0.5235693	0.1481383
512	0.0155359	0.5445120	0.1383758
513	0.0194775	0.5656900	0.1289942
514	0.0239928	0.5869653	0.1200751
515	0.0291000	0.6082000	0.1117000
516	0.0348148	0.6293456	0.1039048
517	0.0411202	0.6503068	0.0966675
518	0.0479850	0.6708752	0.0899827
519	0.0553786	0.6908424	0.0838453

(to be continued)

Wavelength (nm)	$\overline{x}(\lambda)$	$\overline{y}(\lambda)$	$\overline{z}(\lambda)$
520	0.0632700	0.7100000	0.0782500
521	0.0716350	0.7281852	0.0732090
522	0.0804622	0.7454636	0.0686782
523	0.0897400	0.7619694	0.0645678
524	0.0994564	0.7778368	0.0607884
525	0.1096000	0.7932000	0.0572500
526	0.1201674	0.8081104	0.0539044
527	0.1311145	0.8224962	0.0507466
528	0.1423679	0.8363068	0.0477528
529	0.1538542	0.8494916	0.0448986
530	0.1655000	0.8620000	0.0421600
531	0.1772571	0.8738108	0.0395073
532	0.1891400	0.8849624	0.0369356
533	0.2011694	0.8954936	0.0344584
534	0.2133658	0.9054432	0.0320887
535	0.2257499	0.9148501	0.0298400
536	0.2383209	0.9237348	0.0277118
537	0.2510668	0.9320924	0.0256944
538	0.2639922	0.9399226	0.0237872
539	0.2771017	0.9472252	0.0219892
540	0.2904000	0.9540000	0.0203000
541	0.3038912	0.9602561	0.0187180
542	0.3175726	0.9660074	0.0172404
543	0.3314384	0.9712606	0.0158636
544	0.3454828	0.9760225	0.0145846
545	0.3597000	0.9803000	0.0134000
546	0.3740839	0.9840924	0.0123072
547	0.3886396	0.9874182	0.0113019
548	0.4033784	0.9903128	0.0103779
549	0.4183115	0.9928116	0.0095293
550	0.4334499	0.9949501	0.0087500
551	0.4487953	0.9967108	0.0080352
552	0.4643360	0.9980983	0.0073816
553	0.4800640	0.9991120	0.0067854
554	0.4959713	0.9997482	0.0062428
555	0.5120501	1.0000000	0.0057500
556	0.5282959	0.9998567	0.0053036
557	0.5446916	0.9993046	0.0048998
558	0.5612094	0.9983255	0.0045342
559	0.5778215	0.9968987	0.0042024
560	0.5945000	0.9950000	0.0039000
561	0.6112209	0.9926005	0.0036232
562	0.6279758	0.9897426	0.0033706
563	0.6447602	0.9864444	0.0031414
564	0.6615697	0.9827241	0.0029348
565	0.6784000	0.9786000	0.0027500
566	0.6952392	0.9740837	0.0025852
567	0.7120586	0.9691712	0.0024386
568	0.7288284	0.9638568	0.0023094
569	0.7455188	0.9581349	0.0021968

(to be continued)

(continued)

Wavelength (nm)	$\overline{x}(\lambda)$	$\overline{y}(\lambda)$	$\overline{z}(\lambda)$
570	0.7621000	0.9520000	0.0021000
571	0.7785432	0.9454504	0.0020177
572	0.7943256	0.9384992	0.0019482
573	0.8109264	0.9311628	0.0018898
574	0.8268248	0.9234576	0.0018409
575	0.8425000	0.9154000	0.0018000
576	0.8579325	0.9070064	0.0017663
577	0.8730816	0.8982772	0.0017378
578	0.8878944	0.8892048	0.0017112
579	0.9023181	0.8797816	0.0016831
580	0.9163000	0.8700000	0.0016500
581	0.9297995	0.8598613	0.0016101
582	0.9427984	0.8493920	0.0015644
583	0.9552776	0.8386220	0.0015136
584	0.9672179	0.8275813	0.0014585
585	0.9786000	0.8163000	0.0014000
586	0.9893856	0.8047947	0.0013367
587	0.9995488	0.7930820	0.0012700
588	1.0090892	0.7811920	0.0012050
589	1.0180064	0.7691547	0.0011467
590	1.0263000	0.7570000	0.0011000
591	1.0339827	0.7447541	0.0010688
592	1.0409860	0.7324224	0.0010494
593	1.0471880	0.7200036	0.0010356
594	1.0524667	0.7074965	0.0010212
595	1.0567000	0.6949000	0.0010000
596	1.0597944	0.6822192	0.0009686
597	1.0617992	0.6694716	0.0009299
598	1.0628068	0.6566744	0.0008869
599	1.0629096	0.6438448	0.0008426
600	1.0622000	0.6310000	0.0008000
601	1.0607352	0.6181555	0.0007610
602	1.0584436	0.6053144	0.0007237
603	1.0552244	0.5924756	0.0006859
604	1.0509768	0.5796379	0.0006454
605	1.0450000	0.5668000	0.0006000
606	1.0390369	0.5539611	0.0005479
607	1.0313608	0.5411372	0.0004916
608	1.0226662	0.5283528	0.0004354
609	1.0130477	0.5156323	0.0003835
610	1.0026000	0.5030000	0.0003400
611	0.9913675	0.4904688	0.0003072
612	0.9793314	0.4780304	0.0002832
613	0.9664916	0.4656776	0.0002654
614	0.9528479	0.4534032	0.0002518
615	0.9384000	0.4412000	0.0002400
616	0.9231940	0.4290800	0.0002295
617	0.9072440	0.4170360	0.0002206
618	0.8905020	0.4050320	0.0002120
619	0.8729200	0.3930320	0.0002022

(to be continued)

Wavelength (nm)	$\overline{x}(\lambda)$	$\overline{y}(\lambda)$	$\overline{z}(\lambda)$
620	0.8544499	0.3810000	0.0001900
621	0.8350840	0.3689184	0.0001742
622	0.8149460	0.3568272	0.0001556
623	0.7941860	0.3447768	0.0001360
624	0.7729540	0.3328176	0.0001168
625	0.7514000	0.3210000	0.0001000
626	0.7295836	0.3093381	0.0000861
627	0.7075888	0.2978504	0.0000746
628	0.6856022	0.2865936	0.0000650
629	0.6638104	0.2756245	0.0000569
630	0.6424000	0.2650000	0.0000500
631	0.6215149	0.2547632	0.0000442
632	0.6011138	0.2448896	0.0000395
633	0.5811052	0.2353344	0.0000357
634	0.5613977	0.2260528	0.0000326
635	0.5419000	0.2170000	0.0000300
636	0.5225995	0.2081616	0.0000276
637	0.5035464	0.1995488	0.0000256
638	0.4847436	0.1911552	0.0000236
639	0.4661939	0.1829744	0.0000218
640	0.4479000	0.1750000	0.0000200
641	0.4298613	0.1672235	0.0000181
642	0.4120980	0.1596464	0.0000162
643	0.3946440	0.1522776	0.0000142
644	0.3775333	0.1451259	0.0000121
645	0.3608000	0.1382000	0.0000100
646	0.3444563	0.1315003	0.0000077
647	0.3285168	0.1250248	0.0000054
648	0.3130192	0.1187792	0.0000032
649	0.2980011	0.1127691	0.0000013
650	0.2835000	0.1070000	0.0000000
651	0.2695448	0.1014762	0.0000000
652	0.2561184	0.0961886	0.0000000
653	0.2431896	0.0911230	0.0000000
654	0.2307272	0.0862648	0.0000000
655	0.2187000	0.0816000	0.0000000
656	0.2070971	0.0771206	0.0000000
657	0.1959232	0.0728255	0.0000000
658	0.1851708	0.0687101	0.0000000
659	0.1748323	0.0647698	0.0000000
660	0.1649000	0.0610000	0.0000000
661	0.1553667	0.0573962	0.0000000
662	0.1462300	0.0539550	0.0000000
663	0.1374900	0.0506738	0.0000000
664	0.1291467	0.0475496	0.0000000
665	0.1212000	0.0445800	0.0000000
666	0.1136397	0.0417587	0.0000000
667	0.1064650	0.0390850	0.0000000
668	0.0996904	0.0365638	0.0000000
669	0.0933306	0.0342005	0.0000000

(to be continued)

(continued)

Wavelength (nm)	$\overline{x}(\lambda)$	$\overline{y}(\lambda)$	$\overline{z}(\lambda)$
670	0.0874000	0.0320000	0.0000000
671	0.0819010	0.0299626	0.0000000
672	0.0768043	0.0280766	0.0000000
673	0.0720771	0.0263294	0.0000000
674	0.0676866	0.0247080	0.0000000
675	0.0636000	0.0232000	0.0000000
676	0.0598068	0.0218008	0.0000000
677	0.0562822	0.0205011	0.0000000
678	0.0529710	0.0192811	0.0000000
679	0.0498186	0.0181207	0.0000000
680	0.0467700	0.0170000	0.0000000
681	0.0437840	0.0159038	0.0000000
682	0.0408754	0.0148372	0.0000000
683	0.0380726	0.0138107	0.0000000
684	0.0354046	0.0128348	0.0000000
685	0.0329000	0.0119200	0.0000000
686	0.0305642	0.0110683	0.0000000
687	0.0283806	0.0102734	0.0000000
688	0.0263448	0.0095333	0.0000000
689	0.0244528	0.0088462	0.0000000
690	0.0227000	0.0082100	0.0000000
691	0.0210843	0.0076238	0.0000000
692	0.0195999	0.0070854	0.0000000
693	0.0182373	0.0065915	0.0000000
694	0.0169872	0.0061385	0.0000000
695	0.0158400	0.0057230	0.0000000
696	0.0147906	0.0053430	0.0000000
697	0.0138313	0.0049958	0.0000000
698	0.0129487	0.0046764	0.0000000
699	0.0121292	0.0043801	0.0000000
700	0.0113592	0.0041020	0.0000000
701	0.0106294	0.0038384	0.0000000
702	0.0099388	0.0035891	0.0000000
703	0.0092884	0.0033542	0.0000000
704	0.0086788	0.0031341	0.0000000
705	0.0081109	0.0029290	0.0000000
706	0.0075824	0.0027381	0.0000000
707	0.0070887	0.0025599	0.0000000
708	0.0066273	0.0023932	0.0000000
709	0.0061954	0.0022373	0.0000000
710	0.0057903	0.0020910	0.0000000
711	0.0054098	0.0019536	0.0000000
712	0.0050526	0.0018246	0.0000000
713	0.0047175	0.0017036	0.0000000
714	0.0044035	0.0015902	0.0000000
715	0.0041094	0.0014840	0.0000000
716	0.0038339	0.0013845	0.0000000
717	0.0035757	0.0012913	0.0000000
718	0.0033343	0.0012041	0.0000000
719	0.0031091	0.0011227	0.0000000

(to be continued)

(continued)

Wavelength (nm)	$\overline{x}(\lambda)$	$\overline{y}(\lambda)$	$\overline{z}(\lambda)$
720	0.0028993	0.0010470	0.0000000
721	0.0027043	0.0009766	0.0000000
722	0.0025230	0.0009111	0.0000000
723	0.0023542	0.0008501	0.0000000
724	0.0021966	0.0007932	0.0000000
725	0.0020492	0.0007400	0.0000000
726	0.0019110	0.0006901	0.0000000
727	0.0017814	0.0006433	0.0000000
728	0.0016601	0.0005995	0.0000000
729	0.0015464	0.0005584	0.0000000
730	0.0014400	0.0005200	0.0000000
731	0.0013400	0.0004839	0.0000000
732	0.0012463	0.0004500	0.0000000
733	0.0011585	0.0004183	0.0000000
734	0.0010764	0.0003887	0.0000000
735	0.0009999	0.0003611	0.0000000
736	0.0009287	0.0003354	0.0000000
737	0.0008624	0.0003114	0.0000000
738	0.0008008	0.0002892	0.0000000
739	0.0007434	0.0002684	0.0000000
740	0.0006901	0.0002492	0.0000000
741	0.0006405	0.0002313	0.0000000
742	0.0005945	0.0002147	0.0000000
743	0.0005519	0.0001993	0.0000000
744	0.0005124	0.0001850	0.0000000
745	0.0004760	0.0001719	0.0000000
746	0.0004424	0.0001598	0.0000000
747	0.0004115	0.0001486	0.0000000
748	0.0003830	0.0001383	0.0000000
749	0.0003566	0.0001288	0.0000000
750	0.0003323	0.0001200	0.0000000
751	0.0003098	0.0001118	0.0000000
752	0.0002889	0.0001043	0.0000000
753	0.0002695	0.0000973	0.0000000
754	0.0002516	0.0000908	0.0000000
755	0.0002348	0.0000848	0.0000000
756	0.0002192	0.0000791	0.0000000
757	0.0002045	0.0000738	0.0000000
758	0.0001908	0.0000689	0.0000000
759	0.0001780	0.0000643	0.0000000
760	0.0001662	0.0000600	0.0000000
761	0.0001550	0.0000560	0.0000000
762	0.0001446	0.0000522	0.0000000
763	0.0001349	0.0000487	0.0000000
764	0.0001258	0.0000454	0.0000000
765	0.0001174	0.0000424	0.0000000
766	0.0001096	0.0000396	0.0000000
767	0.0001022	0.0000369	0.0000000
768	0.0000954	0.0000344	0.0000000
769	0.0000890	0.0000321	0.0000000

Appendix 3

Data of spectral reflectance curves R(Λ) at 10nm wavelength intervals for the 8 color samples used in the calculation of CIE color rendering index

Wavelength (nm)	#1	#2	#3	#4	#5	#6	#7	#8
380	0.219	0.070	0.065	0.074	0.295	0.151	0.378	0.104
390	0.252	0.089	0.070	0.093	0.310	0.265	0.524	0.170
400	0.256	0.111	0.073	0.116	0.313	0.410	0.551	0.319
410	0.252	0.118	0.074	0.124	0.319	0.492	0.559	0.462
420	0.244	0.121	0.074	0.128	0.326	0.517	0.561	0.490
430	0.237	0.122	0.073	0.135	0.334	0.531	0.556	0.482
440	0.230	0.123	0.073	0.144	0.346	0.544	0.544	0.462
450	0.225	0.127	0.074	0.161	0.360	0.556	0.522	0.439
460	0.220	0.131	0.077	0.186	0.381	0.544	0.488	0.413
470	0.216	0.138	0.085	0.229	0.403	0.541	0.448	0.382
480	0.214	0.150	0.109	0.281	0.415	0.519	0.408	0.352
490	0.216	0.174	0.148	0.332	0.419	0.488	0.363	0.325
500	0.223	0.207	0.198	0.370	0.413	0.450	0.324	0.299
510	0.226	0.242	0.241	0.390	0.403	0.414	0.301	0.283
520	0.225	0.260	0.278	0.395	0.389	0.377	0.283	0.270
530	0.227	0.267	0.339	0.385	0.372	0.341	0.265	0.256
540	0.236	0.272	0.392	0.367	0.353	0.309	0.257	0.250
550	0.253	0.282	0.400	0.341	0.331	0.279	0.259	0.254
560	0.272	0.299	0.380	0.312	0.308	0.253	0.260	0.264
570	0.298	0.322	0.349	0.280	0.284	0.234	0.256	0.272
580	0.341	0.335	0.315	0.247	0.260	0.225	0.254	0.278
590	0.390	0.341	0.285	0.214	0.232	0.221	0.270	0.295
600	0.424	0.342	0.264	0.185	0.210	0.220	0.302	0.348
610	0.442	0.342	0.252	0.169	0.194	0.220	0.344	0.434
620	0.450	0.341	0.241	0.160	0.185	0.223	0.377	0.528
630	0.451	0.339	0.229	0.154	0.180	0.233	0.400	0.604
640	0.451	0.338	0.220	0.151	0.176	0.244	0.420	0.648
650	0.450	0.336	0.216	0.148	0.175	0.258	0.438	0.676
660	0.451	0.334	0.219	0.148	0.175	0.268	0.452	0.693
670	0.453	0.332	0.230	0.151	0.180	0.278	0.462	0.705
680	0.455	0.331	0.251	0.158	0.186	0.283	0.468	0.712
690	0.458	0.329	0.288	0.165	0.192	0.291	0.473	0.717
700	0.462	0.328	0.340	0.170	0.199	0.302	0.483	0.721
710	0.464	0.326	0.390	0.170	0.199	0.325	0.496	0.719
720	0.466	0.324	0.431	0.166	0.196	0.351	0.511	0.725
730	0.466	0.324	0.460	0.164	0.195	0.376	0.525	0.729
740	0.467	0.322	0.481	0.168	0.197	0.401	0.539	0.730
750	0.467	0.320	0.493	0.177	0.203	0.425	0.533	0.730
760	0.467	0.316	0.500	0.185	0.208	0.447	0.565	0.730

References

1 Boyce PR. Human Factors in Lighting. London: Applied Science, 1981.

2 Commission Internationale de l'Eclairage. CIE Publication 13.2, Method of Measuring and Specifying Colour Rendering Properties of Light Sources. Paris: CIE, 1974.

3 Commission Internationale de l'Eclairage. CIE Publication 15.2, Colorimetry. Vienna: CIE, 1986.

4 Loe DL, Mansfield KP, Rowlands E. A step in quantifying the appearance of a lit scene. Lighting Research & Technology 2000; 32: 213-222.

5 MacAdam DL. Maximum visual efficiencies of colored materials. Journal of the Optical Society of America 1935; 25: 361-367.

6 MacAdam DL. Color Measurement. Berlin: Springer-Verlag, 1981.

7 Ohta N, Wyszecki G. Designing illuminants that render given objects in prescribed colors. Journal of the Optical Society of America 1976; 66: 269-275.

8 Xu H. A study of the Colour Rendering Capacity of a light source. Lighting Research & Technology 1983; 15: 185-189.

9 Xu H. Color rendering capacity of illumination. Journal of the Optical Society of America 1983; 73: 1709-1713.

10 Xu H. Colour rendering capacity and luminous efficiency of a spectrum. Lighting Research & Technology 1993; 25: 131-132.

11 Xu H. Lightness and chroma of computer simulated surfaces lit by lamps of different spectra. Lighting Research & Technology 2002; 34: 289-295.

12 Xu H. Correspondence: Colour rendering capacity. Lighting Research & Technology 2008; 40: 257.

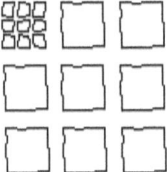

Color Rendering Capacity of Light

To assess how many colors on earth can be vividly rendered by
a given light source, or to see whether a given light source is
capable of rendering a great number of widely different colors,
a measure called the Color Rendering Capacity has been
devised. This book introduces the idea of color rendering
capacity of light, explains the procedure for calculating the
color rendering capacity of any light source, after reviewing
some of the most important concepts in color science.

Xu He (H.Xu) was a pupil in the YaoHua School of Tianjin, a
graduate of Physics from NanKai University, a postgraduate
tackling a project entitled "The function of illumination in terms
of communication theory" in the Bartlett School of University
College London, and an engineer with a research institute in
Beijing before writing this book.

www.ingramcontent.com/pod-product-compliance
Lightning Source LLC
Chambersburg PA
CBHW021019180526
45163CB00005B/2031